Canguilhem

Key Contemporary Thinkers series includes:

Canguilhem

Stuart Elden

polity

First published in 2019 by Polity Press

Polity Press
65 Bridge Street
Cambridge CB2 1UR, UK

Polity Press
101 Station Landing
Suite 300
Medford, MA 02155, USA

ISBN-13: 978-1-5095-2877-6
ISBN-13: 978-1-5095-2878-3 (pb)

A catalogue record for this book is available from the British Library.

Library of Congress Cataloging-in-Publication Data

Names: Elden, Stuart, 1971- author.
Title: Canguilhem / Stuart Elden.
Description: Medford, MA : Polity, 2019. | Series: Key contemporary thinkers | Includes bibliographical references and index.
Identifiers: LCCN 2018037564 (print) | LCCN 2018054119 (ebook) | ISBN 9781509528813 (Epub) | ISBN 9781509528776 (hardback) | ISBN 9781509528783 (pbk.)
Subjects: LCSH: Canguilhem, Georges, 1904–1995.
Classification: LCC B2430.C355 (ebook) | LCC B2430.C355 E43 2019 (print) | DDC 194–dc23
LC record available at https://lccn.loc.gov/2018037564

Typeset in 10.5 on 12 pt Palatino
by Toppan Best-set Premedia Limited
Printed and bound in the UK by TJ International Limited

For further information on Polity, visit our website: politybooks.com

Contents

Acknowledgements

This book was written in parallel to my ongoing research on Foucault for Polity Press, especially for the forthcoming *The Early Foucault*. For this project, I am grateful to friends and colleagues for assistance, encouragement and suggestions: Giuseppe Blanco, Stefanos Geroulanos, G. M. Goshgarian, Inanna Hamati-Ataya, Mark Kelly, Daniele Lorenzini, Gerald Moore, Nicolae Morar, Ingrid Muller, Clare O'Farrell, Simon Reid-Henry, Alison Ross, and Couze Venn. I also thank Pascal Porcheron, Ellen MacDonald-Kramer, John Thompson and their colleagues at Polity for their enthusiasm for this project, and the anonymous readers of the proposal and manuscript. Leigh Mueller copyedited the manuscript, and Lisa Scholey compiled the index.

Much of the initial research for this book was conducted while I was a visiting scholar at ACCESS Europe at the University of Amsterdam in 2017. I thank Luisa Bialasiewicz for arranging the visit, and Guy Geltner for the use of his office. I have consulted materials at the following libraries: University of Amsterdam, Bibliothèque Nationale de France, Bibliothèque Sainte-Geneviève, Beinecke Rare Book and Manuscript Library at Yale University, British Library Rare Books Room and Newsroom, Columbia University, London School of Economics, Senate House Library, University College London, University of Warwick and the Wellcome Library. I am especially grateful to Nathalie Queyroux and David Denéchaud at the Centre d'Archives en Philosophie, Histoire et Édition des Sciences (CAPHÉS) at the École Normale Supérieure in Paris for access to the Georges Canguilhem archive of his papers and library.

Thanks, as ever, to Susan for her love and support.

Abbreviations for Works by Georges Canguilhem

In-text references are provided for major works. For books, the French page number is given first, followed by the English after a slash. A dash means the relevant text does not have a corresponding passage. I have sometimes modified existing translations, especially earlier ones, for clarity and consistency.

Throughout, English titles are used for works available in translation; French for untranslated texts or unpublished manuscripts, though an English translation of the title is provided the first time they are used. Greek characters are transliterated.

BT   'Le cerveau et la pensée', in *Georges Canguilhem: Philosophe, historien des sciences – Actes du colloque (6-7-8 décembre 1990)*, Paris: Albin Michel, 1993, 11–33; 'The Brain and Thought', trans. Steven Corcoran and Peter Hallward, *Radical Philosophy* 148, 2008, 7–18

DE   *Du développement à l'évolution au XIXe siècle*, with Georges Lapassade, Jacques Piquemal and Jacques Ulmann, Paris: Presses Universitaires de France, 2003 [1962]

EGC   'Entretien avec Georges Canguilhem', in François Bing, Jean-François Braunstein, and Elisabeth Roudinesco (eds.), *Actualité de Georges Canguilhem: Le normal et le pathologique*, Paris: Synthélabo, 1998, 121–35

EHPS   *Études d'histoire et de philosophie des sciences*, Paris: Vrin, 5th edn, 1983 [1968]. Includes an additional study, while the pagination for the rest replicates the first edition

FCR   *La formation du concept de réflexe aux XVIIe et XVIIIe siècles*, Paris: Presses Universitaires de France, 2nd edn, 1977 [1955]

IR   *Idéologie et rationalité dans l'histoire des sciences de la vie: Nouvelles études d'histoire et de philosophie des sciences*, Paris: Vrin, 1977; trans. Arthur Goldhammer, *Ideology and Rationality in the History of the Life Sciences*, Cambridge, MA: MIT Press, 1988. Subsequent French editions have the 1977 pagination in the margins

KL   *La connaissance de la vie*, Paris: Vrin, 2nd revised edn, 1965 [1952]; trans. Stefanos Geroulanos and Daniela Ginsburg, *Knowledge of Life*, New York: Fordham University Press, 2009. Subsequent French editions have the 1965 pagination in the margins

NP   *Le normal et le pathologique*, Paris: Presses Universitaires de France, 12th edn, 2015 [1943/1966]; trans. Carolyn R. Fawcett and Robert S. Cohen as *The Normal and the Pathological*, New York: Zone, 1991 [1978]

OC I   *Oeuvres complètes tome I: Écrits philosophiques et politiques (1926–1939)*, ed. Jean-François Braunstein and Yves Schwarz, Paris: Vrin, 2011

OC IV   *Oeuvres complètes tome IV: Résistance, philosophie biologique et histoire des sciences 1940–1965*, edited by Camille Limoges, Paris: Vrin, 2015

RAM   'The Role of Analogies and Models in Biological Discovery', in A. C. Crombie (ed.), *Scientific Change*, London: Heinemann, 1963, 507–20; French version in EHPS 305–18

VM   *Vie et mort de Jean Cavaillès*, Paris: Allia, 1996
VR   *A Vital Rationalist: Selected Writings*, edited by François Delaporte, translated by Arthur Goldhammer, New York: Zone, 1994

WM   *Écrits sur la médecine*, Paris: Seuil, 2002; trans. Stefanos Geroulanos and Todd Meyers, *Writings on Medicine*, Fordham University Press, 2011

WP   'What is Psychology?', trans. Howard Davies, *Ideology and Consciousness* 7, 1980, 37–50; French version in EHPS 365–81

Archival Material

CAPHÉS   Archives de Georges Canguilhem, Centre d'Archives en Philosophie, Histoire et Édition des Sciences (CAPHÉS), École Normale Supérieure.

1

Foundations

Preface

Georges Canguilhem was a significant twentieth-century thinker, usually described as a historian and philosopher of science. He wrote extensively on politics, medicine, biology, history and epistemology. Some of his work concentrates on the formation of specific medical and biological concepts, such as the reflex, bacteria, evolution, regulation and psychology. There are also historical studies of science, including essays on figures such as Gaston Bachelard, Claude Bernard, Jean Cavaillès, Auguste Comte and Charles Darwin. His relation to Bachelard is especially important, and he develops Bachelard's research on physics and mathematics to apply his ideas to the life sciences, as well as developing his epistemological claims, particularly concerning obstacles and ruptures. Canguilhem edited some of Bachelard's work for publication, and regularly wrote prefaces to other people's works.[1] In his work on biology, he developed accounts of milieu and experience, or examined themes in natural history. For Canguilhem, these broad questions have important political and social consequences: they relate to concrete human problems.

He published five single-authored books in his lifetime: *The Normal and the Pathological* (1943), *Knowledge of Life* (1952), *La formation du concept de réflexe aux XVIIe et XVIIIe siècles* [The formation of the concept of reflex in the seventeenth and eighteenth centuries] (1955), *Études d'histoire et de philosophie des sciences* [Studies in the history and philosophy of the sciences] (1968) and *Ideology and Rationality*

(1977).[2] Some of the earlier works included additional essays when re-edited for later editions. He also led a collaborative project in his Paris seminars entitled *Du développement à l'évolution au XIXe siècle* [From development to evolution in the nineteenth century], first published in 1962.

His most famous book is his first, *The Normal and the Pathological*, written in 1943 as his doctoral thesis in medicine, and revised in 1966 with some additional essays. His doctoral thesis in philosophy was *La formation du concept de réflexe*, and for this degree *Knowledge of Life* was also submitted as the minor thesis. In a sense, *La formation du concept de réflexe* and *The Normal and the Pathological* are the only books he wrote. All his other books are collections of essays or lectures. A posthumous collection, *Writings on Medicine* (2002), brought together some late essays, and a six-volume *Oeuvres complètes* [Complete works] is under way. This is an invaluable project, since many other articles, chapters and prefaces first appeared in a diverse range of outlets, which are often hard to find.

This study is an introduction to his work as a whole, drawing on the entirety of his writings, including the *Oeuvres complètes*, and to some extent on archival material housed at the Centre d'Archives en Philosophie, Histoire et Édition des Sciences (CAPHÉS) at the École Normale Supérieure (ENS) in Paris. This archive holds all of Canguilhem's surviving papers, as well as his personal library. While, therefore, it is thoroughly researched in terms of his familiar and more obscure writings, its aim is to make Canguilhem's important and sometimes difficult ideas accessible to as wide an audience as possible.

The book is organized in a thematic and partly chronological way. Chapter 1, 'Foundations', examines Canguilhem's life and career as a whole, especially the years of his intellectual formation. Chapter 2 discusses *The Normal and the Pathological*, both in its original and revised form, and some later essays on its themes. This remains Canguilhem's most influential work, and it established his status as a major thinker. Yet Canguilhem's work was broader than this study alone would suggest.

Chapter 3 focuses on a series of early lectures on philosophical biology, which set out many of the topics he was to discuss in future work. These lectures discussed the questions of vitalism and mechanism in relation to living beings, and the situation of organisms within a milieu or environment. Their importance to his overall career cannot be overestimated. They were published, along with some more specific studies, in *Knowledge of Life* in 1952. In later writings, Canguilhem utilized some of these insights, along with his keen historical sense, to examine a number of questions in the history of the life sciences. Chapter 4 looks at the question of physiology, focusing on the role of experimentation in science, and his major study of the concept of the reflex. Chapter 5 explores some of his other historical studies on regulation and psychology. Chapter 6 looks at his work on evolution and monstrosity. These three chapters draw on both his book-length study *La formation du concept de réflexe* and the *Knowledge of Life* collection, but also on his multiple essays, many of which were collected in *Études d'histoire*.

Yet Canguilhem was significant not just as a historian, but as a philosopher of history, through his discussions of knowledge, truth and ideology. These themes come through in many places, notably *Études d'histoire* and its sequel *Ideology and Rationality*. They are the focus of chapter 7, which also discusses his debt to Cavaillès and Bachelard. Chapter 8 discusses some later essays on medicine, many of which were collected in the posthumous book *Writings on Medicine*. Finally, chapter 9, 'Legacies', discusses the influence his work has had, on both some of his contemporaries and his students, as well as more broadly.

Life and career

Canguilhem was born on 4 June 1904 in Castelnaudary in the Occitania region of southern France. His father owned a smallholding in the commune, and both parents were from rural families, but his father was also a tailor (EGC 132). Canguilhem's occasional claim that he was a peasant is therefore only partially true.[3] Nonetheless, all his childhood friends became farmers, and this is what he was expected to become.[4] Something of the limits of his upbringing can be understood from Canguilhem telling Pierre Bourdieu that when he first attended school he did not know what the toilets were for.[5]

He was taught at the Lycée Henri IV in Paris (1921–4), and then entered the École Normale Supérieure (ENS) in 1924 to study philosophy, where he was an exact contemporary of Raymond Aron, Daniel Lagache, Paul Nizan and Jean-Paul Sartre.[6] Jean Cavaillès had entered the previous year, while Jean Hyppolite (1925), Maurice Merleau-Ponty (1926) and Simone Weil (1928) were also students at a similar time.[7] Bourdieu reports that Canguilhem felt a distance from his friends Aron and Sartre in terms of their attitude to sports – he played the team game of rugby; they were tennis players.[8] His

Diplôme d'études supérieures thesis, roughly equivalent to a Master's degree, was written between November 1925 and June 1926 and directed by Célestin Bouglé. The topic was 'the theory of order and progress in Auguste Comte'.[9] Even at this very early stage of his career, Canguilhem was interested in the relation between philosophical reflection on science, the specific nature of the life sciences, and the way a historical approach could help to elucidate their concerns.

In the *agrégation de philosophie* in 1927, he came second, after Paul Vignaux and above Jean Lacroix and Cavaillès.[10] While teaching was the next academic step, he first had to do his compulsory military service. Influenced in part by the philosopher Alain (Émile Chartier), who had taught him at Henri IV, and also by the atmosphere in the ENS, Canguilhem was a pacifist, and had been active in student circles on this question.[11] It was expected that he would enter the ENS officer training programme but he avoided this by dropping a machine gun tripod on the foot of an inspecting officer. His service between November 1927 and April 1929 was as a private, later promoted to corporal. Following the completion of his service, Canguilhem taught philosophy in *lycées* in Charleville (1929–31), Douai (1932–3), Valenciennes (1933–5), Béziers (1935–6) and Toulouse (1936–40).[12]

In 1935, Canguilhem was the author of an anonymously published pamphlet on *Le fascisme et les paysans* [Fascism and the peasants], published under the auspices of the Comité de Vigilance des Intellectuels Antifascistes.[13] As part of a series of pamphlets under the general title of Vigilance, this group also presented reports on *Qu'est-ce que le Fascisme?* and *Les croix de feu: leur chef, leur programme* in the same year. Paul Rivet was the president of the Comité; Alain and Paul Langevin were its vice presidents.

While teaching at the town's *lycée*, Canguilhem had taken up training in medicine, initially at the University of Toulouse. He later went on to doctoral studies at the University of Strasbourg. Canguilhem was, therefore, formally trained in both philosophy and medicine, a crucial combination for his later research, though he spent his career as a professor of philosophy. Colin Gordon has suggested that Canguilhem's initial medical training was solely in order to become sufficiently knowledgeable to undertake work in the history of the life sciences.[14] This is partly the way that Canguilhem himself recalls things, though he stresses that it was more to have some practical experience aside from just book-learning (EGC 121–2).

Canguilhem's first major philosophical work was published with Camille Planet in 1939, as *Traité de logique et de morale*.[15] This early period has been the focus of some of the most interesting work recently, including a fundamental study by Xavier Roth.[16] The *Traité* advertised that two further works were forthcoming by the same authors – *Traité de psychologie* and *Traité d'esthétique* – but neither of these was ever published. The Second World War, of course, intervened, and things changed. Following the French surrender in June 1940, Canguilhem refused to continue teaching under the Vichy regime. He said he had 'not taken the *agrégation de philosophie* in order to teach the Vichy regime's insipid morality of 'Work, Family, Fatherland'.[17] Instead he joined the French resistance, in the Maquis band of rural guerrillas, alongside his friend Cavaillès, a significant historian and philosopher of logic and mathematics (EGC 123). Canguilhem operated for the resistance under the codename of 'Lafont', and made unanticipated use of his medical training. For organizing a field hospital in Auvergne, and evacuating it while under attack in the battle of Mont Mouchet, he was awarded the Croix de Guerre and the Médaille de la Résistance.[18]

When Cavaillès moved to the Sorbonne in 1941 to take up the chair in logic, he encouraged Canguilhem to replace him in the philosophy department at the University of Strasbourg (EGC 122).[19] At the time, the faculty were in exile in Clermont-Ferrand because of the German wartime occupation of Alsace-Lorraine. During those war years, Canguilhem researched and wrote his doctoral thesis in medicine, *The Normal and the Pathological*, defended in 1943. In 1948, he became the Inspector General of Philosophy in the French higher education system, a post he had initially turned down immediately after the liberation, preferring to return to Strasbourg. He also became President of the Jury d'Agrégation in philosophy, exercising a major influence on intellectual life.

In 1953, he led a UNESCO report on the teaching of philosophy, to which he contributed an overview chapter on 'The Significance of the Teaching of Philosophy' and one on 'The Teaching of Philosophy in France'.[20] The 1948–55 period was the only time he was not teaching his own courses, but teaching was obviously at the heart of his concerns.[21] In 1955, he defended his doctoral thesis in philosophy, *La formation du concept de réflexe*, which was directed by Gaston Bachelard. That same year, he succeeded Bachelard to the chair of the history and philosophy of the sciences at the Sorbonne, and as Director of the Institut d'histoire des sciences et des techniques (IHS) of the University of Paris. He occupied these

positions until his retirement in 1971, though he continued to pursue
an active research and speaking career.

Canguilhem found the events of May 1968 difficult as they put
him in an awkward position. He was friend and mentor to some
of the activists, yet a staunch defender of the educational establishment. He complained of the destruction of things that had taken
years to build.[22] In 1983, he received the George Sarton Medal of
the History of Science Society, and in 1987 the Médaille d'or of the
Centre National de la Recherche Scientifique (CNRS), the highest
scientific distinction in France. Until his final years, he continued
to lecture, including the opening addresses at two important confer-
ences on Foucault's work held in 1988 and 1991.[23]

His early work is shaped by Alain, Bouglé and Henri Bergson.
Alain remained a lifelong friend, and Canguilhem was there when
he died in 1951. Foucault suggests that Canguilhem is both 'far
from and close to Nietzsche', noting that, while Nietzsche saw 'truth
as the greatest lie', for Canguilhem it was science's 'most recent
error'.[24] Indeed, Canguilhem told Michel Fichant he saw himself as
a 'non card-carrying Nietzschean'.[25] Canguilhem is most commonly
seen as part of a French tradition in the history and philosophy of
science of which Bachelard, Cavaillès and Alexandre Koyré were
key figures.[26] Canguilhem's work is certainly not a systematic philosophy, but nor is it just the work of a historian of science.[27] Indeed,
although his work certainly does trace the histories of some sciences,
he never described himself as a historian in a narrow sense, and
his institutional positions were all in philosophy. Yet there is relatively little traditional philosophy in his published works. Intrigu-
ingly, in 1947, Canguilhem signed a contract to write a book in an
introductory series with Bordas: *Pour connaître la pensée de Descartes*
[To know the thought of Descartes]. Henri Lefebvre wrote the volumes
on Lenin and Marx in that series; there were also volumes on Alain,
Bergson, Nietzsche and Bachelard, among others. Yet, while Canguilhem wrote over half of the text, and submitted some chapters
to the pubisher, he never completed the work.[28]

There is no straightforward ethics in Canguilhem's mature work,
though he did discuss this in his earliest work with Planet. Equally,
The Normal and the Pathological, with its genesis in the war and the
sole period he practised as a physician, shows the political stakes
of his work. His last writings on medicine return to the ethical
duties inherent in this practice. It would be difficult to claim any
of his work was metaphysics, although there are reflections on space,
time and mortality, to name just a few themes. He wrote little that

could be considered aesthetics, though his books and lectures are occasionally illustrated with examples from art and literature. A course on the analogue and the singular is introduced by Honoré de Balzac's *La peau de chagrin*;[29] he mentions Francisco Goya's paintings and the poem *Orlando Furioso* (KL 178/140, 181/143) and begins an essay on Darwin with Gustave Flaubert's *Bouvard and Pécuchet* (EHPS 112). He seems content to have left his contribution to logic to just his early work with Planet.

In December 1990, a conference on his work was convened at the Palais de la Découverte science museum in Paris. Canguilhem sent a letter to the organizers, politely declining to attend due his age – he was 86 at the time. But the letter is also revealing because he tells them that 'It is not possible, at my age, to do other than what I have always done, that is to consider what is called my *oeuvre* as anything other than the trace of my trade [*métier*]'. He added that he was well aware of might seem to be the 'rustic nature of his behaviour'.[30] His teaching was indeed crucial to his sense of vocation. The majority of Canguilhem's publications relate to his teaching in some way. But there is not a sense of his research directing his teaching as much as the other way round. He taught a course at Strasbourg on 'Norms and the Normal' in 1942–3, at the time he was completing his thesis on the normal and the pathological.[31] He taught a course on the 'Normal and Pathological, Norm and Normal' in 1962–3 at the Sorbonne, and one on 'The Normal and the Pathological' the following year, around the time he was writing some additional essays for the book's re-edition.[32] Most of the pieces that make up his various collections of essays can be traced to teaching or guest lectures. Many of those seem to have been him using material from his Paris teaching for an audience elsewhere. For example, he taught a course in Paris on the 'History of Teratology from Étienne Geoffroy Saint-Hilaire' in 1961–2, gave a lecture in Brussels on 'Monstrosity and the Monstrous' on 9 February 1962, and included this lecture in the second edition of *Knowledge of Life*.[33] Courses on scientific and medical ideology in the late 1960s and early 1970s fed into the book published as *Ideology and Rationality*.[34] Yet, equally, there are a number of courses taught which were never developed into publications. At the very least, there was depth of engagement and breadth of interests in his teaching, in which he gave courses on new topics nearly every year, which are only partially reflected in his writings.

Canguilhem, unlike many of his contemporaries, shunned the limelight of media attention. He gave very few interviews, wrote little for newspapers and was not a regular presence on French radio or television.[35] Bourdieu recalls this with regret that he left that space 'to show-offs and impostors', but also with praise for his being true to himself.[36] One rare interview was conducted just a few months before his death on 11 September 1995 in Marly-le-Roi, a small town to the west of Paris where he had lived for many years. Canguilhem's voice mainly comes to us from the traces of his teaching.

Canguilhem had a reputation of being intellectually ferocious, with reports that he believed he could correct erroneous ideas by shouting at people.[37] Bourdieu recalls 'his gravelly voice and accent, which had the effect of making him seem always angry, and the sidelong glance coupled with an ironic smile which accompanied his stern judgements on the academic world'.[38] In his role at the IHS he organized many events, proceedings of some of which have been published. His active participation in the discussions at these events can be seen in some of these publications. In a colloquium on the theme of mathematics and informal doctrines, for example, his exchange with Jacques Guillerme contains the following: 'I found your examples very good, and well analysed. What I see less well is the relation between the examples and the initial exposition of the doctrine for, on the one hand, the informal and, on the other, the relation of all this to the problem of this colloquium ... The question I will ask you is a bit brutal ...'[39] After a somewhat evasive response from Guillerme, he adds: 'That is not really my question: My question was the localization of the informal.' He concludes: 'I did not say that you were mistaken on the situation of the informal. What I asked you was where you placed the informal, because for me I do not see clearly where you place it.'[40] It seems that Canguilhem was suggesting that his work was 'not even wrong'.

However, his sarcasm and violent temper were usually directed at colleagues, rather than students, and he was much loved by those he taught.[41] His nicknames of 'le Cang' or 'King Cang' stems from them – as well as giving a sense of his status, it is a clue to the pronunciation of his name, since it sounds the same as the cinematic ape. These students include a wide array of figures, many now more famous than him. In some ways, he has generally been recognized as influential, rather than for his own work.

It was through his role as president of the Jury d'Agrégation that he first met Foucault, with whom he was to have a lifelong professional relationship (EGC 126). He was the sponsor of Foucault's thesis on madness in 1961 – 'supervisor' would be too strong a word, as Canguilhem himself makes clear.[42] Hyppolite suggested Foucault contact Canguilhem about his manuscript, and Canguilhem was so impressed he replied: 'don't change anything, it is a thesis'.[43] Despite Foucault's fulsome acknowledgement in the published version,[44] Canguilhem denies that his comments on it made any difference to its final form.[45] Canguilhem later recalled that 'as it happened, I had previously reflected and written on the normal and pathological. Reading Foucault fascinated me while revealing to me my limits'.[46] His importance to Foucault is certainly significant, though Canguilhem was also influenced by Foucault in turn, and the supplementary essays in the revised edition of *The Normal and the Pathological* bear the influence of Foucault's *History of Madness* and *Birth of the Clinic*.[47] There are many links between their projects. As early as 1957, Canguilhem had suggested that 'an archaeology of science is an enterprise which has a meaning, a prehistory of science is an absurdity' (OC IV, 731).

Louis Althusser stated that his 'debt to Canguilhem was incalculable',[48] while Canguilhem's work on science and ideology, especially in the 1970s, also builds on Althusser (EGC 128). Many of Althusser's students attended Canguilhem's seminars, and the work of the collaborative *Reading Capital* bears the mark of these discussions. This is noted by Étienne Balibar in his preface to the work, in which he and his colleagues pay tribute to Canguilhem (along with Bachelard, Cavaillès and Foucault) as one of their 'masters in reading learned works'.[49] The influence is also felt in the journal *Cahiers pour l'analyse*. Appearing in ten issues between 1966 and 1969, the *Cahiers* was edited by some of Althusser's students at the ENS, including Badiou, Jean-Claude Milner and Jacques-Alain Miller.[50] While drawing on a range of intellectual inspirations, including Althusser and Lacan, the *Cahiers* gave Canguilhem a privileged position, reprinting his lecture 'What is Psychology?' in issue 2, and with a passage from one of his essays on Bachelard used as an epigraph for each issue: 'To work on a concept is to move between extension and comprehension, to generalise it through the incorporation of the marks of exception, to take it outside of its original region, to take it as a model or inversely to search for its model, in short, gradually to confer on it, through regulated transformations, the function of a form' (EHPS 206).[51]

Canguilhem supervised Balibar's, Dominique Lecourt's and Pierre Macherey's Master's theses, and Bourdieu began a doctoral thesis under his supervision.[52] Jacques Derrida was nominally his assistant at an early stage of his career, and Canguilhem was instrumental in getting Derrida's Introduction to his translation of Edmund Husserl's *The Origin of Geometry* the Jean Cavaillès prize. Canguilhem and Derrida kept up a friendship and correspondence until the end of Canguilhem's life.[53] The legacies of Canguilhem's work will be explored more fully in chapter 9.

Oeuvres complètes, English translations and secondary literature

Since his death, and despite Canguilhem's challenge to the idea of his work constituting a coherent whole, there has been an ambitious plan to collect his writings in a multi-volume *Oeuvres complètes*. Two volumes have been published to date, with a third forthcoming and three more planned. Together these will contain almost all the shorter pieces he produced in his long career. They comprise a collection of his early political and philosophical writings (volume I), his work from the 1940s to the 1960s on resistance, the philosophy of biology and the history of sciences (volume IV), and shorter writings from the last three decades of his career on the history of sciences, epistemology and commemorations (the much-delayed volume V). The first of these volumes makes available the anonymous 1935 pamphlet *Le Fascisme et les paysans* and the hard-to-find 1939 book *Traité de logique et de morale*. Both these texts indicate his early interests were rather broader than later specialisms. Many of his early publications were published under pseudonyms, especially 'C. G. Bernard', and these are now integrated into his chronology. Some of his early concerns lead directly to topics addressed in volume IV of the *Oeuvres complètes*, which demonstrates his political engagement from the resistance to the German occupation through to opposition to the Algerian war and General de Gaulle. This volume and the fifth, though, with their extensive collection of texts on the history of sciences, also show how his early work developed into his better-known concerns. Each of these volumes extends to over 1,000 pages. In particular, these volumes make Canguilhem's philosophical underpinnings and political inspirations more explicit.

The planned volume II will contain the three theses *The Normal and the Pathological*, *Knowledge of Life* and *La formation du concept de réflexe*. Volume III will include *Études d'histoire*, *Ideology and Rationality* and the collaborative *Du développement à l'évolution*. The final volume VI will comprise an annotated bibliography, a biography, an index across all the volumes, and 'certain administrative texts and *agrégation* reports'.[54] The texts from the posthumous collection *Writings on Medicine* are in volumes IV and V in their chronological place, while the texts included in other collections, reprinted in volumes II and III, are merely noted in the chronological order of the other volumes. While the forthcoming volumes will be invaluable, the present study is able to draw on almost the entirety of their contents. The books and collections reprinted in volumes II and III have been referenced in their separate editions. Perhaps the key missing element that this book would have wished for was the biography slated to appear in volume VI. That was also something Canguilhem resisted: he told the organizers of the 1990 colloquium that he did not want a biography in the volume that came from it, stating 'I am not dead yet'.[55]

Canguilhem has been sporadically translated into English. *The Normal and the Pathological* was translated in 1978, *Ideology and Rationality in the History of the Life Sciences* in 1988, and the collection *A Vital Rationalist* appeared in 1994, shortly before his death. This collection excerpted parts of the earlier translations, along with parts of other books, and some other writings. But, as David Macey notes, the collection is flawed because, instead of providing full essays or chapters, it comprises 'edited extracts arranged in thematic order. Sentences and even whole paragraphs have been cut and there is nothing to bring the elisions to the reader's attention.' The collection suffers too from 'the complete abolition of chronology', and this 'makes it impossible to trace the development of Canguilhem's thought ... Canguilhem's work was always characterized by a scrupulous attention to detail: King Cang deserves better than this.'[56] In the last decade, two further complete books have appeared in translation – *Knowledge of Life* and the posthumous collection *Writings on Medicine*. His study of the reflex awaits a full translation, and many of his essays on the history and philosophy of science and natural history are not yet available. As far as I am aware, there are currently no plans to translate the rich material of the *Oeuvres complètes*. Nonetheless, this work of translation has helped to feed a growing interest in Canguilhem, for both those with an interest in his substantive topics and those working on the figures he influenced.

The secondary literature in English can be found in a range of places, including the introductions to his books, single chapters in broader studies, and journal articles. The most extensive discussions are in books by Gary Gutting and Dominique Lecourt.[57] Both are

excellent studies, but Gutting discusses Canguilhem in one long
chapter alongside Bachelard as a background to Foucault, and
Lecourt's analysis is only a short part of his study, most of which
is also devoted to Bachelard. Both these books were published while
Canguilhem was still alive and working, and so do not account for
all his work. There are valuable but short introductions in several
of the English collections – by Foucault in *The Normal and the Pathological*, by Paul Rabinow in *A Vital Rationalist*, Stefanos Geroulanos
and Todd Meyers in *Writings on Medicine*, and Meyers and Paola
Marrati in *Knowledge of Life*. Canguilhem is usefully discussed in
studies by Alain Badiou, Bourdieu, Roberto Esposito, Mike Gane,
Hans-Jörg Rheinberger, and Élisabeth Roudinesco, many of which
are available in English.[58] All these are relatively brief discussions
in single chapters of books which treat several thinkers.[59] Given his
importance, a study of the entirety of his work in English is overdue.[60]
    In French, there is significantly more work, including Lecourt's
introductory study in the popular 'Que sais-je?' series. There are
also book-length studies by François Dagognet, Macherey, Claude
Debru, and two by Guillaume Le Blanc.[61] Other monographs con-
centrate explicitly on his work on the normal and pathological, or
on health and illness,[62] and there are important studies of his epistemology.[63] There are also several collections of essays on his work
in French.[64] There is a literature in other languages, notably Spanish,
Italian and German. There are, of course, several articles on his
work in English and French, usually with a specific focus. A special
issue of *Revue de métaphysique et de morale* was devoted to his work
in 1985, including one of the last pieces Foucault revised for publication, a bibliography and a useful chronology of his teaching.[65]
Prospective et santé discussed his work in 1986/7, and *Economy and
Society* in 1998, with contributions by Lecourt, Macey, Rabinow,
Gordon and others. There were also special issues of *Revue d'histoire
des sciences* in 2000, and a bilingual one of *Dialogue: Canadian Philosophical Review* looked at his work in 2013.[66] Since 2007, *Les Cahiers
du Centre Georges Canguilhem* has published seven volumes.
    Much of the difficulty in approaching Canguilhem's work comes
not from his own concepts and writing, but from the material with
which he is grappling. As Lecourt usefully summarizes:

> Canguilhem's texts are undoubtedly disconcerting. The tightly-knit
> style, with its sentences entirely mustered around the concepts which
> give them their order, leaving no room for the slightest rhetorical
> 'play', is rarely reminiscent of what is customary for philosophical

discourse. It does not invite reverie, it does not even urge meditation: it *demands* of the reader that he set himself to work. Nor is there any doubt that the precision of the references and dates, the profusion of proper names, disappoint the expectations of the 'enlightened amateur', half absent-minded, half dilettante, that the philosopher reading a book by one of his peers imagines himself to be, in function if not by right.

There will be more readiness to applaud the erudition than to reflect on the theoretical import of this superabundance of precision.[67]

As this book demonstrates, Canguilhem is a historically crucial figure, bridging an older tradition through to post-structuralism, connecting to debates in Marxism and social theory, while being a thinker who has much to offer to contemporary concerns. Despite Lecourt's comments about precision, it is worth noting that, in common with many other French writers of his generation and the one that followed, Canguilhem can be careless with his references. Sometimes none are provided, and others are incomplete or inaccurate. Fortunately, his editors and translators have corrected many of these. In addition, many of his sources are not available in English. I have rechecked all his references, and have indicated English sources if these exist.[68]

It is not a case here of trying to reconstruct a project which Canguilhem himself did not quite produce, in the sense of a systematization of his work.[69] Rather, its aim is to try to outline what his work did actually accomplish, with the intention that this book can serve as a guide, summary and introduction to his writings. In his study of the reflex, Canguilhem commented: 'A library annexed to a laboratory is eventually divided into two sections: a museum and a workshop. There are books that you glance over like you observe a flint axe; there are others you slice [*dépouille*] like you use a microtome. Where is the boundary between the museum and the workshop? Who traces it, and when does it move?' (FCR 156).

Part of the point of this study is to rescue his work from the museum.

2

The Normal and the Pathological

The Normal and the Pathological remains Canguilhem's best-known work. He outlines his biographical route to this thesis in medicine: studies in philosophy followed by studies in medicine, 'parallel to teaching philosophy' (NP 7/33). He suggests that philosophers can be interested in medicine for reasons that go beyond mental illness or to 'exercise scientific discipline'. Rather, medicine was a means of access to 'concrete human problems' (NP 8/34). The book was reissued several times. In 1950, he says that the budget did not allow more than a reprint, though he suggests additions and corrections were needed (NP 3/29). He confines himself, there, to a few indications of additional reading, though he notes that in 1943 it was difficult for books to circulate (NP 3–4/29). The most significant revision was in the 1966 version, which added three essays to the initial thesis but again left the main body unchanged. As he notes in the foreword to the 1966 edition, the work was 'fortunate enough to arouse interest in medical as well as philosophical circles' (NP i/25). The re-edition came at a time when the original book was receiving renewed attention – including through some of Althusser's students and the work of the *Cahiers pour l'analyse* – and, through being read anew, was connected to a series of debates in the social sciences and humanities.[1]

Canguilhem begins by suggesting that, while 'the problem of pathological structures and behaviours in man is enormous', and poses 'innumerable questions which, in the end, refer to the whole of anatomical, embryological, physiological and psychological research', the question is not helped by a fragmentary, specialized approach. Instead, he suggests that more clarity will come if the problem is 'considered as a whole than if it is broken down into question of detail'. However, he suggests that he is not yet in a position to be able to do this with sufficient documentation, but that this might be a future project (NP 7/33). Perhaps the successive editions of this text, especially with its 1966 additions, along with two separate essays on the theme, were his attempt to fulfil that promise. Part of the problem is that 'medicine seemed to us and still seems to us like a technique or art at the crossroads of several sciences, rather than, strictly speaking, as a science' (NP 8/34).

He suggests that 'the relations between science and technique, and that of norms and the normal' would benefit from an examination of medicine. As such, he describes the work as 'an effort to integrate some of the methods and attainments of medicine into philosophical speculation' (NP 8/34). But he cautions that this is not to try to improve medicine by normative judgement or importing a metaphysics. That is a task for physicians – and it is telling that, even with this thesis in hand, Canguilhem does not self-identify in that way. Rather, the purpose is the other way round: medical knowledge can supplement and develop methodological and philosophical questions (NP 8/34). But equally revealing, here, is his resistance to the idea that he is trying to write 'the work of an historian of medicine. If we have placed a problem in historical perspective in the first part of our book, it is only for reasons of greater intelligibility. We claim no erudition in biography' (NP 9/34).

The project orientates itself around a generally adopted thesis from the nineteenth century, by which 'pathological phenomena are identical to corresponding normal phenomena save for quantitative variations' (NP 9/35). That thesis is subjected to critical examination here, following a philosophical approach Canguilhem claims is derived from Léon Brunschvicg. This is to reopen problems rather than close them, the view that 'philosophy is the science of solved problems: we are making this simple and profound definition our own' (NP 9/35). The work comprises three main parts – the two parts of the first section, written in 1943, and the 1966 additions. The reading here begins with the second part, which sets out key concerns, before returning more briefly to the more historical first part. It then discusses his later thoughts on the topic.

The sciences of the normal and pathological

Part Two begins with the observation that, while psychiatrists have thought and rethought the question of the normal and pathological,

physicians and physiologists have not thought about this in terms
of their work (NP 91/115). Canguilhem is writing this as a thesis
in medicine, and so he is interested in opening up this question of
the 'lived experience of the sick'. He suggests that it is impossible
for the physician to understand this through discussion: 'what the
sick express in ordinary concepts is not directly their experience
but their interpretation of an experience for which they have been
deprived of adequate concepts' (NP 91/115). One of those key concepts is what normal actually is, what being healthy really means.
Canguilhem says that he agrees with René Leriche's claim that 'health
is life lived in the silence of the organs',[2] and that, as a consequence,
what we call 'biologically normal' is 'revealed only through infrac-
tions of the norm and that concrete or scientific awareness of life
exists only through disease' (NP 94/118).

Canguilhem gives the example of a man who cut his arm badly
on a circular saw. This man will recover much use of the arm, but
compared to his other arm it will not be normal, even if he is able
to do much again with it (NP 95–6/119–20). He also discusses a
farmhand whose leg was broken and rehealed badly (due to the
neglect of his employer), and who had his leg rebroken and set: 'It
is clear that the [medical] department head who made the decision
had another image of the human leg than that of the poor devil
and his master.' Canguilhem adds that the norm being adopted
would not have satisfied an Olympic runner or a ballet dancer (NP
97/121).

The question of the normal then is mutable, and adapted to context.
He thinks that physicians have tended to avoid it because it seems
'too vulgar or too metaphysical'. They are interested in 'diagnosis
and cure', but both of those implicitly rely on a notion of a norm
from which the organism has deviated. They take their understanding of this norm from three sources: 'knowledge of physiology –
called the science of the normal man – from his lived experience of
organic functions, and from the common representation of the norm
in a social milieu at a given moment'. He suggests that the first is
the most significant, because 'modern physiology is presented as a
canonical collection of functional constants related to the hormonal
and nervous functions of regulation. These constants are termed
normal insofar as they designate average characteristics, which are
most frequently practically observable' (NP 98/122).

The question of the relation between the norm and the normative
arises here. 'Physiological constants are thus normal in the statistical
sense, which is a descriptive sense, and in the therapeutic sense,

which is a normative sense.' Canguilhem is interested in how medicine situates itself in relation to these two aspects of the question, in whether medicine converts 'descriptive and purely theoretical concepts into biological ideals', or whether medicine accepts 'the notion of norm in the normative sense of the word' (NP 98–9/122–3).

Canguilhem turns to dictionary definitions in search of a sense of the normal. A medical dictionary simply says that the normal is 'that which conforms to the rule, regular'. A philosophical dictionary is more helpful: *norma* 'etymologically, means a T-square [*équerre*], normal is that which bends neither to the right nor left, hence that which remains in a happy medium [*milieu*]' (NP 101/125, see NP 227/239).[3] This, in turn, leads to two meanings: '(1) normal is that which is such that it ought to be; (2) normal, in the most usual sense of the word, is that which is met with in the majority of cases of determined kind, or that which constitutes either the average or standard of a measurable characteristic' (NP 101/125). The ambiguity comes, Canguilhem underlines, from the way that the term seems to be both a fact, derived statistically, and a value attributed to that fact. This may be compounded in philosophical realism, where generalities are taken to be essences, and perfection as their realization; and so 'a generality observable in fact takes the value of realized perfection, and a common characteristic, the value of an ideal type' (NP 101/125). In medicine, this tension is apparent too, where the normal state is both the habitual and the ideal, where the aim of therapeutics is to restore organs to the habitual (NP 101–2/126).

In medicine, the aim is to re-establish the 'normal state of the human body'. But Canguilhem asks whether it is 'because therapeutics aims at this state as a good to obtain that it is called normal, or is it because the interested party – that is the sick person – considers it normal that therapeutics aim at it?' He believes that the second is true, that 'life is a normative activity' (NP 102/126). Canguilhem claims that the notion of the normative in philosophy combines the sense of a fact and a judgement relative to it, but 'this mode of judgment is essentially subordinate to that which establishes norms. Normative, in the fullest sense of the word, is that which establishes norms. And it is in this sense that we plan to talk about biological normativity' (NP 102–3/127). The relation of the norm to the pathological, and the abnormal, is also significant. The pathological is a more tightly circumscribed notion: 'Biological pathology exists but there is no physical or chemical or mechanical pathology' (NP 103/127). Nor is the pathological the simple opposite of the normal: 'There is no biological indifference, and consequently we can speak of biological normativity. There are healthy biological norms and there are pathological norms, and the second are not the same as the first' (NP 105/129). Indeed, as he goes on to suggest, 'it is life itself and not medical judgment which makes the biological normal a concept of value and not a concept of statistical reality' (NP 107/131).

The English nouns 'anomaly' and 'abnormality' are partnered by the adjectives 'anomalous' and 'abnormal', but the situation in French is somewhat different. Drawing on André Lalonde's *Vocabulaire philosophique*, Canguilhem notes that the noun *anomalie* and the adjective *anormal* are distinct words which had become fused – in English this is 'anomaly' and 'abnormal'. As he explains: '*Anomalie* is a substantive with no corresponding adjective at present; *anormal*, on the other hand, is an adjective with no substantive, so that [French] usage has coupled them, making *anormal* the adjective of *anomalie*' (NP 107/131).[4]

He goes on to note that there is a French word *anomal*, anomalous, but this has 'fallen into disuse' (NP 107/131). Canguilhem stresses that the suffix of these words is not related to *nomos*, the Greek word for law, but to *omalos*, what is level or smooth: '"*anomalie*" is etymologically, *an-omalos*, that which is uneven, rough, irregular, in the sense given these words when speaking of a terrain' (NP 107/131).[5] He also underlines that we must be careful not to conflate the Greek *nomos* and the Latin *norma*, even though they have 'closely related meanings, law and rule tending to be confused' (NP 107–8/132). The other crucial point which arises from these etymological and semantic issues is that some of these terms are descriptive and others evaluative. Anomaly is, he suggests, a fact, a description of a state of affairs. Abnormality suggests a value judgement, a normative approach. These distinct senses have, however, become confused, in part because of the imprecision of grammatical understanding: '"Abnormal" has become a descriptive concept and "anomaly" a normative one' (NP 108/132).

This semantic confusion is widespread, Canguilhem argues, and can be found in the work of Isidore Geoffroy Saint-Hilaire – and, following him, even some medical dictionaries. But this does not mean that Saint-Hilaire's work on the notion can be dismissed, and it is actually significant to understanding how the term is used. Canguilhem notes that, for Saint-Hilaire, it is wrong, concerning animals, to speak of either *'peculiarities [bizarreries] of nature ... or disorder or irregularity ...* If there is an exception, it is to the laws of naturalists, not the laws of nature, for in nature all species are *what they must be* in this grand *ensemble* where throughout there is, in a famous expression, variety in unity and unity in variety.'[6] Anomaly is therefore morphological: it is some 'unusual, unaccustomed' removal from 'the vast majority of beings to which one must be compared'. It then relates to '*the specific type* and *individual variation*' (NP 109/133). 'It is clear that, so defined, anomaly is, generally speaking, a purely empirical or descriptive concept, a statistical deviation' (NP 109/133).

Canguilhem therefore asks whether anomaly and monstrosity are equivalent. For Saint-Hilaire, 'monstrosity is one species of the genus anomaly. Whence the division of anomalies into *Varieties, Structural defects, Heterotaxy* and *Monstrosities*' (NP 109/133). Summarizing Canguilhem's summary, these are progressively more extreme, though not in a simple linear sense. The first are minor, cosmetic deviations; the second are deformations of a more significant kind, making one or more functions impossible; the third are complex structural shifts that nonetheless do not impede function; the final ones are both complex in appearance and serious in impairment (NP 109–10/133–4). This is why the linear scale is not sufficient – there is more of a sense of a two-by-two table – simple–complex and slight–serious distinctions being the two axes (NP 110/134). The third is perhaps the most intriguing, since *heterotaxy* is Saint-Hilaire's term, and this phenomenon was rarely studied. It might pertain to an internal structural organization, which exhibited neither functional disorder nor external appearance. As Canguilhem stresses, 'it is difficult to imagine the possibility of a complex anomaly which not only does not obstruct the smallest function but also does not even produce the slightest deformity' (NP 111/135).

Biological sciences therefore need to address questions of anomaly and teratology. Canguilhem stresses that 'not all anomalies are pathological', but when they are identified they have led to a science which studies them. The scientific approach has tended to marginalize the normative implications of this, but has also shaped the way in which we now think of anomaly, which tends to generate ideas of harmful or even mortal deformities, rather than mere statistical divergencies. Additionally, forms of life or behaviours of the living being have been given a normative value, instead of being seen as a statistical norm (NP 112–13/136–7).

The anomaly in itself is therefore not pathological. It is a variation, and this can be measured with the range of statistical deviation. This diversity is not always disease, and 'pathology' implies a sense of 'suffering and impotence', deriving from the term *pathos*. However, the pathological is abnormal. Yet we might say that the pathological is normal too, if we look at statistical deviation, and recognize that 'continual perfect health is abnormal' (NP 113/137). This, however, only arises because of a confusion in the sense of the term 'health'. The notion of absolute health is itself a normative concept, and 'it is a pleonasm to speak of good health because health is organic well-being'. A more useful term is 'qualified health', which 'is a descriptive concept, defining an individual organism's particular disposition and reaction with regard to possible diseases' (NP 114/137). Absolute health is normative and qualified health is descriptive, and the two senses must not be confused: the idea that perfect, continuous health is abnormal is true in the sense of it being unobserved. However Canguilhem is looking at the abnormal in a different sense, where 'the concepts of sick, pathological and abnormal' can be equated (NP 114/138).

The milieu

Chapter 3 will discuss the notion of the milieu in much more detail. But there is some useful discussion here in relation to abnormality. The milieu is the environment, the context, of an organism. A milieu is normal in relation to that organism, allowing it to live there more or less effectively. Canguilhem takes the example of colour variation in butterflies between black and grey. There are three examples of milieux – nature, industrial regions and captivity. In captivity, black butterflies are stronger and eliminate the greys. In nature, black butterflies are more visible against trees and are eaten by birds, whereas the grey can hide better. But in industrial regions, where there are fewer birds, 'butterflies can be black with impunity' (NP 119/143).[7] As Canguilhem summarizes: 'taken separately the living being and its milieu are not normal: it is their relationship that makes them such' (NP 120/143). A living being can adapt to the milieu and thereby appear as normal – a response to that milieu. Anomaly is not pathological because it is an anomaly, because some of these mutations allow the living being to be normal – that is, adaptive. 'In biology the normal is not so much the old as the new form, if it finds conditions of existence in which it will appear normative, that is, displacing all withered, obsolete and perhaps soon to be extinct forms [*formes passées, dépassées ... trépassées*]' (NP 120/144).[8] This is a point that Canguilhem continually stresses:

'There is no fact which is normal or pathological in itself.' What we call an anomaly or mutation is only pathological if it means that there is a limitation of some kind – of terms of life, survival, breeding or some other kind of parameter. The phenomenon is not pathological in itself, in some absolute sense. The phenomenon might be equal or even superior in a different milieu, and, in that setting, normal. 'The pathological is not the absence of a biological norm: it is another norm but one which is, comparatively speaking, pushed aside by life' (NP 121/144).

The example of the butterflies in captivity is therefore revealing. Laboratory work allows the study of normal phenomena in a continuous form, which might lead to the idea that 'there is one possible definition of the normal, objective and absolute, starting from which every deviation beyond certain limits would logically be assessed as pathological' (NP 121/145). But this would be to confuse the laboratory with the outside, and to fail to recognize the norms established through experimentation. Physiology, like physics and chemistry, depends on experiments which produce results that can be compared, 'all other things being equal': 'In other words, other conditions would give rise to other norms. *The living being's functional norms* as examined in the laboratory are meaningful only within the framework of the *scientist's operative norms*' (NP 121–2/145). As a consequence, physiology can give content to biological norms, but not 'work out in what way such a concept is normative' (NP 122/145). A scientist would have to claim that the conditions of experiment do not influence the result in order to claim otherwise, and that goes against the basic principle of care scientists apply to experimentation (NP 122/145).

Indeed, by any external criteria, the laboratory is a pathological situation, in that it is unusual – or at least a statistical divergence from the usual milieu of the living being. There is therefore an irony in using that setting as the context for the examination of the living being, to try to derive an understanding of what is normal, and the statistical weight of a norm (NP 122/146). Canguilhem quotes Victor Prus to the effect that experimental physiology is an artificial pathology,[9] and says that this means that the praise of 'contemporary theorists of wave mechanics for their discovery that observation interferes with the observed phenomenon' is misplaced: the idea is much older than them (NP 123–4/146–7).

Bernard thought that this could be controlled for in experiments,[10] but Canguilhem notes that this is not simple and lists three criteria: the 'normal' subject in the experiment must be the same in a non-artificial situation; the pathological state created in experiment must be the same as the spontaneous one; and the scientist must compare the result of the previous two criteria. He suggests that 'no one will question the breadth of the margin of uncertainty introduced by such comparisons', but adds that 'it is as vain to deny the existence of this margin as it is childish to question *a priori* the utility of such comparisons' (NP 124/147–8). However, the canonical experimental requirement of 'all other things being equal' is, he notes, 'very hard to attain', and it must be recognized that, 'for the animal or for man the laboratory milieu is one possible milieu among others' (NP 125/149).

Norm and average

The statistical aspect of the norm has already been noted, but Canguilhem devotes a whole chapter to this question. The average or mean (*moyenne*) is now often taken to be the basis for the norm in physiology. This was resisted by Bernard because he recognized 'the essentially oscillatory and rhythmic character of the functional biological phenomena', and suggested that 'the normal is defined as an ideal type in determined experimental conditions rather than as an arithmetical average or statistical frequency' (NP 128/152). Canguilhem suggests that the work of Pierre Vendryès is significant in showing how the question became one of divergences from a mean, so that 'the terms divergence [*écart*] and average here have a probabilistic meaning. The greater the divergences the more improbable they are' (NP 128/153).[11] Problems remain with this, of course, because we still need to know 'within what range of oscillations around a purely theoretical average value individuals will be considered normal' (NP 129/154).

While some might want to separate the norm and the average entirely, Canguilhem points to the notion of biometrics, in which research in physiology sees norm and average as 'inseparable concepts', and, because average can be defined objectively, it is natural to want to join norm to it, even though there are 'insurmountable' difficulties in doing this (NP 131/156). Canguilhem's example comes from Adolphe Quetelet's work on 'anthropometric procedures'. If we take the height of men, for example, the more measurements we make, the more extreme outliers will cancel each other out in deriving an average which is also the median point of a bell-shaped probability distribution. Given a large population, '*those who come closest to the average height are the most numerous*; those who diverge from it the most are the least numerous'. For Quetelet, 'the average man is by no means an "impossible man"' (NP 133/157).[12] Canguilhem suggests that 'the interest of Quetelet's conception lies in the fact that in his notion of true average he identifies the ideas of *statistical frequency* and *norm*, for an average which determines that the greatest divergences are the most rare is really a norm' (NP 133–4/158).

Quetelet recognized the differences between a simple midpoint in a range, the median and the average as the mean of all values in the range. He was criticized for importing an ontological regularity instead of an empirical basis for his determination. Part of the reason is that something like height is not simply a biological factor – especially not for a diverse group such as humans – but is 'inseparably biological and social'. Canguilhem notes that 'even if height is a function of the milieu, the product of human activity must be seen, in a sense, in the geographical milieu. Man is a geographical agent and geography is thoroughly penetrated by history in the form of collective technologies' (NP 135/159). It follows from this, he suggests, that 'in the human species, statistical frequency expresses not only vital but also social normativity', a situation which is even more true for a 'physiological characteristic like longevity, rather than an anatomical characteristic' (NP 135/160). A good example of this is age, with the average life span a social rather than biological norm. The average expresses what has become the norm, rather than providing a measure outside of that specific context, a conclusion Canguilhem suggests would be clearer if we looked at subsections of the society in terms of class or occupation, rather than as a whole (NP 137/161).

The geographical dimensions of this are also important to Canguilhem, although he notes that comparative human physiology had not been written at the time. He adds that, as he was completing the study, he was made aware of Maximilien Sorre's work on biology and geography (NP 139/163).[13] He says that if we derive averages from the laboratory, 'one would run the risk of presenting normal man as a mediocre man, far below the physiological possibilities of which men, acting directly and concretely on themselves or the milieu, are obviously capable' (NP 139/164). There is some discussion of physiology and sports records, as well as of yogis and health, and diseases caused by behaviour. But the key issue is the question of ecology, the relation of the human to milieu, and adaptation to it. This is why, he says, 'a geographer's work is of great interest for a methodological essay on biological norms' (NP 145/170).[14]

Canguilhem gives various examples, which are often quite technical. He suggests that the geographical needs to be combined with historical study, though notes that 'paleopathology has even fewer documents at its disposal than palaeontology or palaeography', but suggests that it can be seen in 'every deviation from the healthy state of the body which has left a visible trace on the fossilized skeleton. If the sharpened flints and art of Stone Age men tell the story of their struggles, their works and their thought, "their bones call to mind the history of their pains"' (NP 147–8/172).[15] Again, there are complications. Canguilhem contends that 'the relationship between the biological norms of life and the human milieu seems to be both cause and effect of men's structure and behaviour' (NP 149/173). And we could diagnose all the problems of a skeleton in terms of the norms of today only 'if we were to ignore the differences of cosmic milieu, technical equipment and way of life which make the abnormal of today the normal of yesterday' (NP 149/174). The work of geographers such as Sorre and Paul Vidal de La Blache 'has shown that there is no geographical destiny. Milieux offer man only potentialities for technical utilization and collective activity' (NP 150/175).

One of the aspects he finds interesting is the relationship of humans to location. He notes that Xavier Bichat said that 'animals inhabit the world while plants belong only to the place where their life was born. This idea is even truer of men than of animals', and that, with the exception of spiders, humans have expanded over a greater expanse of the earth. Even more, humans have developed technologies to allow them to change their milieu, showing that they are 'the only species capable of variation'. As such, 'the milieu of the living being is also the work of the living being who chooses to shield himself from or submit himself to certain influences' (NP 153–4/178–9).

Summarizing this discussion, Canguilhem suggests that 'the concepts of norm and average must be considered as two different concepts: it seems vain to try to reduce them to one by wiping out the originality of the first. It seems to us that physiology has better to do than to search for an objective definition of the normal, and that is to recognize the original normative character of life' (NP 153/177–8). One of the more significant tasks of physiology – and no small challenge – 'would then be to determine exactly the content of the norms to which life has succeeded in fixing itself without prejudicing the possibility or impossibility of eventually correcting these norms' (NP 153/178).

Disease, cure, health

In this discussion, Canguilhem makes a number of claims about medicine and biology that he would expand upon in subsequent work, discussed in the other chapters of this book. He is particularly interested in the work of Kurt Goldstein, both his book *The Organism* and his earlier work on head wounds, carried out during the First World War.[16] Canguilhem suggests that 'in distinguishing anomaly from the pathological state, biological variety from negative vital value, we have, on the whole, delegated the responsibility for perceiving the onset of disease to the living being itself, considered in its dynamic polarity' (NP 155/181). Now, drawing on Goldstein, he underlines that the reference must always be the individual, and their relation to the milieu, because what is possible for one individual may not be for others: 'a statistically obtained average does not allow us to decide whether the individual before us is normal or not. We cannot start from it in order to discharge our medical duty toward the individual' (NP 155/181).[17]

This is important, because it moves from the collective, which can be assessed by the calculative, statistical measures, to the individual. As Canguilhem states, the social division of the normal and the pathological is problematic for multiple reasons, if we analyse multiple individuals together, but it can be used to analyse an individual at different time periods. There the boundary may be precise, and transition effectively tracked. But this too is dependent on context: what is normal at one point can be pathological in another setting, or vice versa, if other factors remain the same (NP 156/182).

This, he suggests, is a long way from the positions of Comte or Bernard. It means that the pathological and health are not polar opposites, and that the pathological cannot simply be called 'abnormal' in an absolute sense. It can be seen as abnormal if it varies from specific, but not general, terms (NP 170–1/196–7). He suggests that 'disease is not a variation on the dimension of health; it is a new dimension of life' (NP 160/186). Again, he insists on the importance of the milieu for this determination, and that 'health is a margin of tolerance for the inconstancies of the milieu', though it may be absurd to speak of the last part (NP 171/197). Health is a combination of 'securities in the present', and 'assurances for the future', against these variations; it is a 'regulatory flywheel', whereas 'disease is characterised by the fact that it is a reduction in the margin of tolerance for the milieu's inconstancies' (NP 172–3/198–9).

The issues explored here help us to understand the notion of physiology. Canguilhem cautions against seeing this as 'the science of the laws or constants of normal life', because of this imprecision around the notion of normal. The normal cannot be objectively measured, and 'the pathological must be understood as one type of normal, as the abnormal is not what is not normal, but what constitutes another normal'. This is not to challenge the idea of physiology as a science, but to recognize that it is easier to specify how it works as a science – 'its search for constants and invariants, its metrical procedures, and its general analytical approach' – than what it is a science of. He suggests it would be better to see it as the science of 'the conditions of health' than the science of 'the normal functions of life' (NP 177/203). 'In short, in order to define physiology, everything depends on one's concept of health' (NP 177–8/203–4).

He discusses various options from the historical literature to specify this, and it is a notion he will return to many times in his career. Provisionally, he suggests that 'defining physiology as the *science of the stabilized modes of life*' is probably the most appropriate. Among the benefits of this decision is the ability to understand the relation between physiology and pathology (NP 179/205). Again he works through various definitions, but suggests that medical teaching which begins with 'the anatomy and physiology of the normal man' (NP 181–2/207) now needs to be supplemented with discussion of immunity and allergy (NP 185/211). His summary is important, stressing that the 'distinction between physiology and pathology has and can only have a clinical significance'. Even though medical practice frequently talks of 'diseased organs, diseased tissues, diseased cells', Canguilhem suggests that it is medically incorrect to do so (NP 197/223). Why does Canguilhem claim this? He notes that different biological theorists have placed disease in different places – Giovanni Battista Morgagni at the level of the organ, Bichat at the tissue, Rudolf Virchow at the cell. But he suggests that 'to look for disease at the level of cells is to confuse the plane of concrete life, where biological polarity distinguishes between health and disease, with the plane of abstract science, where the problem gets a solution' (NP 197–8/223). In contrast to these positions, Canguilhem suggests that it only makes sense when taken in a more general way: 'We suggest that it is as a whole that it can be called sick or not' (NP 198/224).

He suggests that 'in pathology the first word historically speaking and the last word logically speaking comes back to the clinic' (NP 200/226). The French term here is *la clinique*, and as Foucault would explore in much more detail, this means both clinical practice and the hospital. This medical setting is significant for technical expertise and for the way this changes how doctors approach their study. When they talk of 'pathological anatomy, a pathological physiology, a pathological histology, a pathological embryology', Canguilhem is insistent that 'their pathological quality is an import of technical and thereby subjective origin', rather than there being an objective pathology. The phenomenon this studies or treats can be described in objective terms, but the judgement of whether it is pathological or not cannot be made on a 'purely objective criterion'. This is the distinction he has been insisting on: 'Objectively, only varieties or differences can be defined with positive or negative vital values' (NP 200–1/226).

A history of the problem

The question of the pathological state as a quantitative modification of the normal state is the topic of the first part of the book. It is largely historical, and enables him to expose the problematic nature of this claim. He works through some key figures from nineteenth-century thought, notably Comte and Bernard, but also his contemporary Leriche, who then occupied the Collège de France chair once held by Bernard (NP 20–1/46).[18] Canguilhem suggests that 'the impetus behind any ontological theory of disease undoubtedly derives from therapeutic need' (NP 13/39). What has entered a body can leave it; what has been lost can be restored; 'we can hope to conquer disease even if it is the result of a spell, or magic, or possession, we have only to remember that disease happens to man in order not to lose all hope' (NP 13/39).

The detail of these readings is extensive, and a full discussion is impossible. The 1942–3 course engaged with Émile Durkheim, whose *The Rules of Sociological Method* had a chapter devoted to 'Rules for Distinguishing between the Normal and the Pathological',[19] though Durkheim is strangely absent from the book. The point of the history – which is a characteristic of almost all his work after this thesis – is to establish the grounds on which the modern conception could develop. But this is not a teleology, where the past inexorably leads to the present. Canguilhem is interested in the dead ends, detours

and reversals of the historical record. Another reason for the historical work is that medicine so often neglects its own past, even as it
uncritically takes forwards some of its prejudices: 'it is a fact to be
reckoned with that people generally enter medicine completely
ignorant of medical theories, but not without preconceived notions
of many medical concepts' (NP 18/44).

There is a shift today from Greek medicine in Hippocrates, where
disease is dynamic, and where nature, *physis*, 'within man as well
as without, is harmony and equilibrium'. Imbalance is associated
with what is called disease, and by redressing the balance the natural
state can be restored. Disease, however, is not the imbalance itself,
but rather the organism's attempt to get well, to restore the balance.
In this understanding, 'medical technique imitates natural medical
action'. Today, though, the attempt to restore 'the diseased organism
to the desired norm' is delegated to technique, which is partly based
on an assumption that 'we can expect nothing good from nature
itself' (NP 14–15/40–1). Canguilhem calls the Greek model a 'dynamic
or functional theory', and the modern one an 'ontological theory',
and suggests that 'medical thought has never stopped alternating
between these two representations of disease, between these two
kinds of optimism, always finding some good reason for one or the
other attitude in a newly explained pathogenesis' (NP 15/41).

In the modern conception is 'the formation of a theory of the
relations between the normal and the pathological, according to
which the pathological phenomena found in living organisms are
nothing more than quantitative variations, greater or lesser according to corresponding physiological phenomena' (NP 16/42). But
this is distinct from the ontological version; it does not see the ideas
of health and sickness as 'qualitatively opposed, or as forces joined
in struggle' (NP 16/42). In summary, 'disease is no longer the object
of anguish for the healthy man; it has become instead the object of
study for the theorist of health. It is in pathology writ large that we
can unravel the teaching of health' (NP 17/43).

The nineteenth century is, he suggests, the crucial moment when
the understanding of the relation between the normal and the pathological became a dogma, and when the biological and medical stress
on these terms extended into philosophy and psychology (NP 17/43).
This is why he devotes so much space to Comte, via François Broussais, and Bernard, because they are key figures for understanding
this nineteenth-century 'dogma' – they 'really played the role, half
voluntarily, of standard-bearer' (NP 20/46). He does not mean
'dogma' to be a disparaging term, 'but rather to stress its scope and

repercussions' (NP 18/44). He suggests that the influence of Comte and Bernard is profound on philosophy, science and literature in the nineteenth century. Nonetheless, they do not exactly coincide: Comte moves from a study of the pathological to the normal, suggesting that speculative work can also yield insights into the knowledge of the normal. Direct experimentation on humans is difficult, and this approach can be profitable. Bernard, following his work on biological experimentation, takes the opposite approach to the same destination: he moves from the normal to pathological and proposes that the identification of the two can help to remedy the pathological (NP 18/43–4). As Canguilhem summarizes: 'in Comte the assertion of identity remains purely conceptual, while Claude Bernard tries to make this identity precise in a quantitative, numerical interpretation' (NP 18/44).

Canguilhem notes that 'Nietzsche borrowed from Claude Bernard precisely the idea that the pathological is homogeneous with the normal' (NP 19/45). This is in one of Nietzsche's notebooks, where he suggests that 'it is the value of all morbid states that they show us under a magnifying glass certain states that are normal – but not easily visible when normal' (NP 20/45). Nietzsche then quotes Bernard's *Leçons sur la chaleur animale*, though it is not clear it is a quotation.[20] The full passage reads:

> Health and disease are not two essentially different modes as the ancient physicians believed and some practitioners still believe. They should not be made into distinct principles, entities which fight over the living organism and make it the theatre of their struggle. These are obsolete medical ideas. In reality, between these two modes of being, there are only differences of degree: exaggeration, disproportion, discordance of normal phenomena constitute the diseased state. There is no case where disease would have produced new conditions, a complete change of scene, some new and special products.[21]

Canguilhem hopes that this initial outline shows:

> that the thesis whose meaning and importance we are trying to define has not been invented for the sake of the cause. The history of ideas cannot be inevitably superimposed on the history of sciences. But as scientists lead their lives as men in a milieu and social setting [*entourage*] that is not exclusively scientific, the history of sciences cannot neglect the history of ideas. In following a thesis to its logical conclusion, it could be said that the modifications it undergoes in its cultural milieu can reveal its essential meaning.
>
> (NP 20/46)

In Canguilhem's own words, Part One examined 'the historical sources and analysed the logical implications of the principle of pathology, so often still invoked, according to which the morbid state in the living being is only a simple quantitative variation of the physiological phenomena which define the normal state of the corresponding function. We think we have established the narrowness and inadequacy of such a principle' (NP 203/227). Part Two suggested that 'types and functions can be qualified as normal with reference to the dynamic polarity of life. If biological norms exist it is because life, as not only subject to the milieu but also as an institution of its own milieu, thereby posits values not only in the milieu but also in the organism itself. This is what we call biological normativity' (NP 203/227).

While this was a thesis in medicine, Canguilhem's interest in wider philosophical questions is readily apparent. The project is also politically motivated. It was written during the first half of the war, while he was involved in the resistance and after he had left his teaching roles in opposition to Vichy. Roberto Esposito has suggested that 'I would say that nothing about his philosophy is comprehensible outside of this military commitment', and that the stakes of the project can be seen in the insistence on the importance of life beyond its reduction to a mere material form. The political decisions made as a result of deeming some life to be normal and some pathological – which took its extreme form in eugenics and genocide – are interrogated through a political examination of how such distinctions are made. Esposito rightly sees here and in later essays on medicine an explicit critique of 'Nazi state medicine, which had made that bio-economic procedure the hinge of its own politics of life and death'.[22] For Geroulanos, there is an explicit politics to Canguilhem's rejection of 'the equivalences between norm, the normal and the healthy and any sense of a given type or norm, national or universal'.[23] His political action is partnered by his intellectual commitment.

Canguilhem's final thoughts in the original thesis are aphoristic and generative. He suggests that 'the pathological state can be called normal to the extent that it expresses a relationship to life's normativity' but 'this normal could not be termed identical to the normal physiological state because we are dealing with other norms' (NP 203–4/227–8). He disputes the idea that the abnormal should be understood as 'the absence of normality', and insists that 'the morbid state is always a certain mode of living' (NP 204/228). Equally, he insists the norm cannot be determined by scientific methods, especially in physiology. Neither the living being nor the milieu alone is normal; it is their relation that allows that judgement. 'Strictly speaking, there is no biological science of the normal. There is a science of biological situations and conditions *called* normal. That science is physiology' (NP 204/228). Medicine finds its meaning in the relation of the living being to its milieu, and its attempts to dominate it, and this is 'why medicine, without being a science itself, uses the results of all the sciences in the service of the norms of life' (NP 205/228–9). In the 1950 preface, he indicates that he felt the book left 'the philosophical door open', and that his conclusions were brief. His text makes clear that this was intentional, and that his intention was 'to lay the groundwork for a future thesis in philosophy': 'I was aware of having sacrificed enough, if not too much, to the philosophical demon in a thesis in medicine. And so I deliberately gave my conclusions the appearance of propositions which were simply and moderately methodological' (NP 6/32).

New reflections

Canguilhem never really delivered on this idea of a more exclusively philosophical discussion of these issues, and when he did submit a doctoral thesis in philosophy, twelve years after this thesis in medicine, it was on a quite distinct topic. Nonetheless, he would return to the themes of this book at several points in his career. The most significant are the three essays he would add to the 1966 edition. These are partly a product of teaching the material again, but with different emphasis and exploring different paths (NP 221–3, 233–5). Some of his reflections are generated by new arguments, but also by the availability of older materials, notably the posthumous publication of Bernard's *Principes de médecine expérimentale* in 1947 (NP 222/234).[24] There had also been a growth in the use of the terms in disciplines outside his original focus. He did not think he 'posed the problem badly at the time' (NP 224/236), but now needed to take account of 'the meaning of the concepts of norm and normal in the social sciences, sociology, ethnology, economics', which 'involve research which in the end – whether it deals with social types, criteria of maladjustment to the group, consumer needs and behaviour, preference systems – tend toward the question of the relations between normality and generality' (NP 223/235).

   This politically sharpens some of the claims that he makes in the original thesis, especially when he focuses on the idea of normalization, suggesting that 'the norm is not a static or peaceful, but a dynamic and polemical concept' (NP 227/239). This is because a norm is not just derived from an analysis of phenomena, but used to shape them: 'A norm, or rule, is what can be used to right, to square, to straighten [*à dresser, à redresser*]. To set a norm, to normalise [*normer, normaliser*], is to impose a requirement on an existence, a given whose variety, disparity, with regard to the requirement, present themselves as a hostile, even more than an unknown, indeterminant' (NP 227/239).

As he goes on to suggest, 'it is not just the exception which proves the rule as rule, it is the infraction which provides it with the occasion to be rule by making rules'. What we call the 'infraction' gives rise not to the rule (*règle*) but regulation (*régulation*), and the notion of the 'normative' is sparked by such an infraction. In Kantian terms, he suggests that 'the condition of the possibility of rules is but one with the condition of the possibility of the experience of rules. In a situation of irregularity, the experience of rules puts the regulatory function of rules to the test' (NP 230/242).[25]

It is in these later reflections that Canguilhem makes a crucial suggestion about the relation between the normal and abnormal:

> The abnormal, as ab-normal, comes after the definition of the normal, it is its logical negation. However, it is the historical anteriority of the future abnormal which gives rise to a normative intention. The normal is the effect obtained by the execution of the normative project, it is the norm exhibited in the fact. In the relationship of the fact there is then a relationship of exclusion between the normal and the abnormal. But this negation is subordinated to the operation of negation, to the correction summoned up by the abnormality. Consequently it is not paradoxical to say that the abnormal, while logically second, is existentially first.
>
> (NP 232/243)

This is a claim with which Foucault, among others, would do much. It opens up the possibility of seeing how a relation which seems so simple – the abnormal is a deviation from the norm – actually works in the reverse way, with the norm derived from a cataloguing of those seen as abnormal. Canguilhem himself indicates many possible ways that this might be explored in these additional essays. He returns to etymology, with the Greek *orthos* being seen as equivalent to the Latin *norma*, and shows its links to orthography, orthodoxy and orthopaedics. He adds that 'orthology is grammar in the sense given it by Latin and medieval writers, that is, the regulation of language usage … Grammar furnishes prime material for reflection on norms' (NP 232–3/244). He sees this relation as working in multiple registers: 'In terms of normalization there is no difference between the birth of grammar in France in the seventeenth century and establishment of the metric system at the end of the eighteenth' (NP 233/244). He claims that 'Richelieu, the members of the National Convention and Napoleon Bonaparte are the successive instruments of the same collective demand. It began with grammatical norms and ended with morphological norms of men and horses for national defence, passing through industrial and sanitary norms' (NP 233/244–5). As a note clarifies, this is again in different fields: 'establishment of conscription and the medical examination of conscripts, establishment of national stud-farms and remount depots' (NP 233 n. 3 / 296 n. 14). Further, he suggests that this shift happens at the turn of the eighteenth to nineteenth century, with the end of the *ancien régime*, the revolution and the establishment of the First Empire: 'Between 1759, when the word "normal" appeared, and 1834 when the word "normalized" appeared, a normative class had won the power to identify – a beautiful example of ideological illusion – the function of social norms, whose content it determined, with the use that class made of them' (NP 235/246).

The politics of this is significant, with links between technological, juridical norms and class politics and programmes for normalization and standardization (NP 235–7/246–8). Foucault's studies of the mad, the sick, prisoners, hermaphrodites and the perverse all relate to these themes. Canguilhem himself suggests that there are complicated relations between the social and the biological, with society both organism and machine. Social regulation has parallels with organic regulation, but also with the idea of organization, a designed system for control and order (NP 241/252).

The second essay added in 1966 discusses organic norms in man. It covers some related ground to his broader essays on biology, which will be discussed in other chapters in this book. Canguilhem notes the work of Hans Selye, which had appeared after the writing of the original book (NP 262–4/271–2). Selye's work on stress as a 'state of organic alarm' had been a recognition of the impact of the external world on the human mind, and its sometime inability to deal with perturbations (NP 4/30). His term 'diseases of adaptation' captures this notion.[26] Canguilhem's discussion also returns to themes around the statistical average, and stresses that 'the concept of *normal* in biology is objectively defined in terms of the frequency of the characteristic so qualified' (NP 252/261). It also re-examines some of the issues around the relation between an organism and its milieu, here stressing the importance of adaptation and natural selection as a mechanism. He suggests that 'strictly speaking, a mutationist theory of the origin of species can define the normal only as the temporarily viable' (NP 254/263). These themes are explored in some of his collaborative work in his seminars around the time he wrote these essays (see chapter 6). One of the things that becomes clear, however, is how much of his subsequent work has reinforced his initial claims. He suggests that, despite twenty years of 'reading and reflecting', he has not been led 'to put in question the interpretation proposed then for the biological foundation of the original concepts of biometry' (NP 256/265). Nor does he think he needs to 'profoundly modify our analysis of the relations between the determination of statistical norms and the evaluation of the normality and abnormality of this or that individual divergence' (NP 256/265). The third new essay examines the concept of error in pathology, a theme which will be discussed in more general terms in chapter 7. Canguilhem suggests that he is 'quite persuaded' that 'the knowledge of life, like the knowledge of society, assumes the priority of infraction over regularity' (NP 278/285).

Concluding, Canguilhem recognizes the challenges of his initial thesis, and thinks that it was 'the temerity of youth' that allowed him to think himself 'equal to the task of a study of medical philosophy on norms and the normal' (NP 281/289). In some ways he is right, and he certainly never attempted a project of such scope and comprehensiveness again. As the following chapters will show, with the exception of his thesis in philosophy, which was a more tightly circumscribed study of the reflex, more limited in both topic and historical period, all his other interventions were to take the form of teaching, talks, essays and other shorter interventions. Yet *On the Normal and the Pathological* sets an agenda for both his own career and that of others. It remains an indispensable work.

3

Philosophy of Biology

In February, March and May 1947, Canguilhem delivered three lectures to Jean Wahl's Collège philosophique in Paris on biology.[1] These lectures – 'Aspects of Vitalism', 'Machine and Organism' and 'The Living and its Milieu' – later appeared in his *Knowledge of Life* collection. They mark a significant broadening of his research away from medicine to wider biological concerns. Each lecture is sufficiently important for a detailed discussion to be helpful in outlining his core concerns. Esposito sees these lectures, among other works by Canguilhem, as politically charged. For Esposito, Canguilhem's interrogation of the philosophy of biology is a political counterposition to 'the Nazi's programmatic antiphilosophical biology'.[2] The explicit politics may seem muted, but some indications of these stakes will become clearer.

'Aspects of Vitalism'

The lecture begins with a reflection on the relation between biology and philosophy. What would be meant by 'biological philosophy', what relation does this have to 'philosophical biology', and does this already mean the biology is speculative and fanciful, rather than grounded and empirical? (KL 83/59). The concept of life is both a biological and a philosophical one, and it may be true to say the question of 'what is biology?' – a science or knowledge of life – is already a philosophical question rather than a biological one. This is even more so if the biology in question is, in Canguilhem's

terms, one 'fascinated by the prestige of the physico-chemical sciences, a biology reduced or reducing itself to the role of a satellite
of these sciences' (KL 83/59). Such a biology, Canguilhem contends,
effaces 'the biological object as such' – it devalues 'its specificity'
(KL 83/59). Yet a biology that does take a distinctive, and autonomous, approach to its subject matter – life – and the mode of
approaching that subject matter is frequently qualified as, or accused
of being, vitalist (KL 83/59–60). For good reasons, Canguilhem suggests, biologists have tended to want to avoid such a charge, because
of either the pejorative value of the term, or the extravagances of
work that has embraced the label, or the figures with which it is
associated (KL 83/60).[3]

However, Canguilhem notes that, because of eighteenth-century
developments, 'the term *vitalism* is appropriate for any biology
careful to maintain its independence from the annexationist ambitions of the sciences of matter' (KL 84/60). This is a telling claim,
because it distinguishes the subject of biology – life – from the
subject of the other sciences – matter in its various forms. It also is
a point of contrast between Canguilhem and Bachelard's empirical
work on mathematical physics. In order to interrogate this, Canguilhem makes two moves. The first is to say that, in order to understand how biology has reached the point where it wants to distinguish
itself from the vitalist charge, we need to understand the history of
biology as much as its current instantiations. 'A philosophy that
asks science for clarifications of concepts cannot remain uninterested
in the construction of this very science' (KL 84/60). The history is
not just a tracing of a sequence of stages in the development of the
science, but is essential in the determination of its orientation (KL
84/60). The second is to say that the scientific defence of vitalism
is not his purpose – he leaves that to biologists. He wants, instead,
to understand vitalism as a philosopher. There are sound precedents
for this, and he finds vitalist tendencies in philosophers from Hippocrates and Aristotle to Marx, Engels and Bergson, and in biologists
such as Bichat, Jean Baptiste Lamarck, Bernard, Hans Driesch and
Goldstein (KL 85/61).

His claim is that the 'vitality of vitalism' is the first aspect that
needs to be interrogated. Putting this into question requires an
examination of 'the search for the meaning [*sens*] of the relationship
between life and science in general, life and the science of life more
specifically' (KL 85/61). Looking at the history of biology is, he
says, revealing. Biology has tended to operate with a series of oppo-
sitions, been seen as divided, and at times closer to each of the

poles: 'Mechanism and Vitalism confront one another on the problem
of structures and functions; Discontinuity and Continuity on the
problem of succession of forms; Preformation and Epigenesis on
the problem of the development of a being; Atomicity and Totality
on the problem of individuality' (KL 85/61).[4]

In a sweeping survey of the tradition, Canguilhem traces the
way that what comes to be called vitalism has been utilized, or
criticized: 'Vitalism is the expression of the confidence the living
being has in life, of the self-identity of life within the living human
being conscious of living' (KL 86/62). This is a qualified support
of vitalism, though not its contemporary instantiations, principally
as a means of challenging the violence of mechanism. He suggests
that 'the art of prognosis prevails over that of diagnosis, on which
it depends. It is as important to predict the course of a disease as
it is to determine its cause' (KL 86/62), and that 'vitalism translates
a permanent exigency of life in the living, the self-identity of life
immanent to the living' (KL 86/62). That is why 'mechanist biologists and rationalist philosophers' criticize vitalism's 'nebulousness'
and 'vagueness' (KL 86/62). Canguilhem sets up a contrast with
mechanism, which he stresses comes from *mēchanē*, an engine (*engin*)
or machine, or a ruse or stratagem, though he wonders whether
these meanings are effectively one (KL 87/63).[5] Mechanism sees the
human as a living being separated from life by science, and using
science to try to regain it; vitalism seeks the life within the living
(KL 87/62).

Canguilhem goes into some detail about the problems of mechanism, which will be more fully explored in the second lecture, but
his critique is essentially this: 'If the animal is nothing more than
a machine, and the same holds for the whole of nature, why is so
much human effort expended in order to reduce them to that?' (KL
87/63). It thus becomes clear that, for Canguilhem, vitalism is not
a method for making sense of life, but what he calls an 'exigency'
– that is, a demand or a necessity. Following the work of Emanuel
Rädl, he suggests that it is 'a morality rather than a theory' (KL
87–8/63).[6] Nature is something of which the human is part, and
which is a part of the human – not something separate from the
human as 'a foreign, indefinable object' (KL 88/63). Science has
often taken the latter approach; but Canguilhem urges the first, in
which natural phenomena are seen as filled with 'life, soul and
meaning [*sens*]'. Taking such an approach, which it is clear he does
himself, and which he suggests characterizes a tradition from 'Plato,
Aristotle, Galen, all the men of the Middle Ages, and a large number

of the men of the Renaissance', means that one 'is fundamentally a vitalist' (KL 88/64).

In a telling definition, he suggests that a vitalist is someone 'who is led to meditate on the problems of life more by the contemplation of an egg than by the handling of a winch or an iron bellows' (KL 88/64). However, there is a problem in that some of the terms used by vitalist philosophers and biologists assume the existence of something that they are trying to prove – the vital principle, vital force, entelechy or *hormé* (KL 91/66). Entelechy might be glossed as one of those other terms: it is understood as a guiding principle or final form of something, coming from the Greek *entelekheia*, a complete or perfect goal. Used by Aristotle, in later thought it came to be associated with the soul, or Leibniz's monad. Canguilhem's sketch of the history of the tradition is revealing, because he challenges both an undervaluing of vitalism, and its overvaluing. This might come in a variety of ways, including seeing someone's earlier research as a contribution made possible by vitalism, when their own conversion to this approach only came later. This work may have led to their vitalism, not the reverse. He insists that just because someone became known for vitalism does not mean all their work should carry that label, and we should not think all their previous discoveries are due to that inflection. Canguilhem's example is Driesch, whose work on sea-urchin eggs led him to vitalism and the theory of entelechy; even though he initially conceived his work on these eggs as a project to confirm a theory of developmental mechanics (KL 92/67).

With the undervaluing of vitalism, Canguilhem notes that it was quite possible for vitalists to make major scientific advancements, citing the work of Caspar Friedrich Wolff in founding modern embryology in the eighteenth century, or Karl Ernst von Baer's development of that work in the nineteenth. Equally, he stresses that as many vitalists as mechanists were crucial to the development of cell theory, and that this approach was not seen as slowing down scientific progress (KL 92–3/67–8). In this essay, Canguilhem also hints at an argument he will make in considerably more detail in his study of the concept of the reflex (see chapter 4). Here he claims that, in neurology, the theory of the reflex 'probably owes its formation more to vitalists than to mechanists ... the later mechanization of reflex theory cannot belie its origins' (KL 93/68).[7]

Nonetheless, Canguilhem offers some cautions. He suggests that there are some biologists who are experimental in their earlier careers, advancing science as they go, who in later life engage in 'philosophical speculation' and supplement 'pure biology with philosophical biology' (KL 93–4/68). While they are obviously free to do this if they choose, Canguilhem suggests they are not above reproach

> for seeking to profit from their capacity as biologist on philosophical terrain. The vitalist biologist who turns philosopher of biology thinks he brings a certain capital with him to philosophy, but in reality he brings to it only a land-income [*rentes*], which continually decreases in the market of scientific values – for the simple reason that research, in which he no longer participates, continues to move forward
>
> (KL 94/68)

Such a biologist loses the prestige of the scientist, because he is no longer actively researching, no longer taking part in its 'internal dynamism'. Driesch is, again, his example here. But Canguilhem is insistent they must not be allowed to preserve their prestige in philosophy: 'Philosophy, being an autonomous enterprise of reflection, does not honour any prestige at all, not even that of the scientist, or – even more rightly – that of the ex-scientist' (KL 94/69).

The next caution is that vitalism might be seen as not allowing mechanism and 'the physico-chemical explication of life' time to complete its work, positing an unknowable substance in place of what has not yet been explained (KL 94/69). In this, Canguilhem draws upon Bachelard's notion of an 'epistemological frontier', and suggests that vitalism's suggestion of the limits of mechanism is perhaps an example of a poorly posed question. Bachelard states that 'any absolute frontier proposed to science is the mark of a badly stated problem ... It is to be feared that scientific thought does not retain traces of philosophical limitations ... Oppressive frontiers are illusory frontiers.'[8] Some vitalists – and Canguilhem's example is Bichat – see the 'instability and irregularity' of life as directly opposed to 'the invariability of physical laws' (KL 95/69). Bichat claims that, while physics and chemistry are bound by the same laws, these do not apply to life: 'To say that physiology is the physics of animals is to give an extremely inexact view of it; I would as much say that astronomy is the physiology of the stars.'[9]

In sum, this position leads to the 'philosophical inexcusable fault' for Canguilhem: 'the classical vitalist accepts the insertion of the living organism into a physical milieu to whose laws it constitutes an exception' (KL 95/70). The physico-chemical approach is imperialist in its attempt to co-opt all phenomena within its ambit, but biology's response cannot be to claim there is another empire somewhere inside theirs. Biology cannot claim there are 'enclaves of indetermination, zones of dissonance, or foyers of heresy' within 'the physico-chemical territory', understood as 'the milieu of inertia, of externally determined movements' (KL 95/70). Rather, 'the originality of the biological' must be asserted over the whole of experience, not 'islets of experience'. This leads Canguilhem to his key criticism: 'In the end, classical vitalism sins, paradoxically, only in its excessive modesty, in its reluctance to universalize its conception of experience' (KL 95/70). The project of vitalism needs to be bolder, Canguilhem argues, and to understand 'matter within life, and the science of matter – which is science *tout court* – within the activity of the living' (KL 95/70). Physics and chemistry, the suggestion seems to be, are truer to their underlying assumptions than biology has been: 'Today this determination has led them to recognize the immanence of measuring to the measured, and to see the content of observation protocols as relative to the very act of observation' (KL 96/70).

It is here that the notion of a milieu becomes so important. A *milieu* is an environment, the surrounding context around a living being, but also the centre of that environment – the *mi-lieu*, the mid-place or location. For Canguilhem, this sense of *milieu* as the middle or centre can be found in the sense of an organism living in the middle of its environment or surroundings, but also relates to parts of an organism within the organism as a whole, taking the organism as the environment within which the tissue or organs are situated: 'The milieu in which one looks for the emergence of life only acquires its meaning as milieu in virtue of the operation of the human living being who takes measurements of it, measurements that bear an essential relation to the technical apparatuses and procedures by which they are made' (KL 96/70). Three centuries of this calculative, mathematical and experimental physics has led it back to its beginning. 'Physics is a science of fields, of milieux', which has forced a reflection on the centre, the living being coordinating this milieu, giving it its conditions of existence. Canguilhem stresses that this does not take anything away from physics, in either its objects of study or its determinist approach. Rather, it situates the work of physics and its interpretations within a wider realm (KL 96/70–1). As he notes elsewhere, 'the concept of a milieu was, in the eighteenth century, a concept of mechanics and physics. Its importation into biology, in the nineteenth century, favoured mechanical conceptions of life' (EHPS 160–1).

A passage from a later lecture helps to make sense of this notion. In Jean Giraudoux's 1937 play *Electra*, a beggar finds a squashed hedgehog on the road.[10] The beggar meditates on why this might be. Why do hedgehogs cross roads? This question does not really make biological sense, says Canguilhem:

A road is a product of human technology, one of the elements of the human milieu – but it has no biological value for the hedgehog. Hedgehogs as such do not cross roads: they explore, in their own way, their own hedgehog milieu, on the basis of their alimentary and sexual impulses. On the contrary, it is man-made roads that cross the hedgehog's milieu, his hunting ground and the theatre of his loves, just as they cross the milieux of the rabbit, the lion, or the dragonfly. Now, experimental method – as the etymology of the word *method* [*methodus*, 'across the path'] shows – is also a sort of road that the human biologist traces through the world of the hedgehog, the frog, the fruit fly, the paramecium, or the streptococcus. The use of concepts and intellectual tools forged by that living scientist, the biologist, in order to understand the experience of life proper to the organism is thus at once both inevitable and artificial. We will not conclude from this that experimentation in biology is useless or impossible. Instead, keeping in mind Bernard's formula that 'life is creation', we will say that the knowledge [*connaissance*] of life must take place through unpredictable conversions, as it strives to grasp a becoming whose meaning is never so clearly revealed to our understanding as when it disconcerts it.

(KL 39/22)[11]

A biological view is thus crucial to both scientific man and living man. Canguilhem notes, though, that mechanist and materialist challenges are made to biology in this sense, which he suggests is because they are 'jealous of its methodological and doctrinal autonomy' (KL 96/71). They therefore accuse vitalism of being 'scientifically retrograde ... politically reactionary or counter-revolutionary' (KL 96/71). But this rests on some problematic elisions. Classical vitalism held a belief in a soul – animism – wherein this soul could direct the body. This distinction between a soul and a body is paralleled in non-human animals by the idea of life and the living body. That view, Canguilhem accepts, has philosophical and political problems, because of its links to 'dualist spiritualism' (KL 97/71). More recently – and remember, Canguilhem was giving this lecture in 1947 – vitalist biology had been used by Nazi ideology, with the argument of wholeness and totality applied to the individual organism and social organism, rather than 'individualist, atomist, and mechanist liberalism' (KL 97/72). Driesch is one example of a thinker whose biological work took on political overtones, with his notion of entelechy re-described as 'the Führer of the organism' after 1933 (KL 98/72).[12]

Yet was vitalism at fault, or was it more to do with Driesch himself? Was the Nazi use of Darwinism and natural selection more a fault of the theory or of those who made the move? Keeping with the biological metaphor, Canguilhem asks: 'Is this a question of biology or parasitism of biology?' (KL 98/72). Biology has taken concepts from politics, and it may be that here politics is simply taking them back. Canguilhem suggests that the fundamental issue in the 'exploitation of antimechanist biological concepts by Nazi sociologists … is the problem of the relation between organism and society' (KL 98/72). Yet he says that no biologist could address that question of social order on the basis of biology alone, and, just as biology cannot be used to justify political and economic hierarchy and exploitation, nor could a politics concerned with the collapse of hierarchies claim there are none in nature, just because of a belief in social justice or a classless society (KL 98/72). Even more so, it is worth stressing that the Nazis adopted a range of scientific claims for their purposes. This included the use of genetics 'to justify racist eugenics, techniques of sterilization and artificial insemination', as well as this perversion of Darwinism 'to justify their imperialism and their politics of *Lebensraum*' (KL 98/72). For Canguilhem, the use of tools and arguments for nefarious purposes cannot be a simple critique of those tools and arguments: 'One can no more honestly reproach a biology concerned with its autonomy for having been utilized by Nazism than one can reproach arithmetic and the calculation of compound interests for having been utilized by capitalist bankers or actuaries' (KL 98/72–3).

This might be challenged by suggesting the importance of facilitation, of creating the conditions of possibility for something to happen. Nonetheless, Canguilhem concludes with a rephrased claim that the adoption of these ideas was due to the self-interest of certain biologists, stressing their 'lack of character and philosophical resoluteness' (KL 98/73). He recognizes that there are political overtones to how vitalism is sometimes seen in relation to political crises, and the doubts about the 'efficacy of capitalist institutions'. But even this risks seeing a biological crisis as a political and social one (KL 99/73). Esposito has suggested that, as with *The Normal and the Pathological*, these remarks show his 'resolute opposition to Nazism', which was conceptual as well as biographical.[13] But, as well as the Nazi experience, Canguilhem is also thinking about another political issue of the time, which was not so much to do with the biologizing of politics but the politicizing of biology. This was the debate about the critique of gene theory and the critique of natural selection in the Soviet Union, particularly led by Trofim Lysenko, and known as Lysenkoism.[14] From the 1930s until the 1950s, this dominated Soviet agriculture, and suggested that dialectics meant that Lamarck's notions of inheritance could be engineered for productive purposes. Supported by Stalin, at the time Canguilhem was lecturing in Paris it was politically impossible to criticize this movement within communism, even though its claims for food production were never realized. A year after Canguilhem's lecture, Lysenkoism became the only theory taught in the USSR, and deviations or criticisms were seen as inherently bourgeois. Canguilhem only refers to this obliquely in this lecture, but will return to it in 'The Living and the Milieu'.

Yet his criticisms of this application of dialectics to biology does not mean he is entirely opposed to Marxism. He sees that there is something valuable in a Marxist approach to dialectics in relation to biology, but for different reasons from the traditional way this is seen. For Canguilhem, it is justified because there are capacities within life which rebel 'against its mechanization', the grounds which gave rise to vitalism. He wants to inscribe vitalism here as 'an exigency rather than a doctrine' – as the 'vitality found in life... life's proper spontaneity'. He approvingly cites Bernard's suggestion that 'life is creation' (KL 99/73).[15] This, rather than a predetermined approach to the questions, is what shows that dialectics in biology is a useful approach.

This lecture, compressed though it is, and often indicative rather than fully worked through, is a fundamental text in Canguilhem's work. It outlines the way that he sees the role of biology as a distinctive science, and the importance of philosophical reflection on it. It shows how a historical approach to its questions can help shed light on contemporary issues. It demonstrates the political stakes of science, and the critique of its co-option by both the right and the left. But, perhaps above all, it argues for the importance of the vitalist approach. The question of life (*vie*), and the living being (*vivant*) demand this. As Canguilhem closes the lecture: 'In the end, to do justice to vitalism is simply to give life back to it' (KL 100/74).

'Machine and Organism'

Towards the end of the 'Aspects of Vitalism' lecture, Canguilhem acknowledges that 'it is, nevertheless, easier to denounce mechanism and scientism in biology in words than it is to give up in fact their postulates and the attitudes they compel' (KL 99/73). Canguilhem had underscored that philosophy's interest in science's concepts has to interrogate the historical construction of those concepts, not just as a sequence, but to ascertain the development and orientation of the science (KL 84/60). However, as Canguilhem cautions elsewhere, 'the past of a present day science is not the same thing as that same science in the past' (IR 15/5). It is perhaps no surprise that in his second lecture, 'Machine and Organism', he provides a detailed history of mechanism in biology.[16]

Canguilhem begins by suggesting that the 'mechanical theory of the organism' had long been seen as the basis for biology, even as a 'dogma', but is today being challenged by dialectical materialism, among other approaches (KL 101/75). He says that the story is actually more complicated than this, and that the philosophical stakes are higher than just 'a matter of doctrine and method in biology' (KL 101/75). In a characteristic move, he suggests that one way to approach this is to reverse the direction of inquiry. Most work has used mechanism and its account of 'the structure and function of an already-constructed machine' to explain the 'structure and function of the organism'. Much less common has been the reverse, to understand the construction of the machine on the basis of the organism (KL 101/76). In addition, both scientists and philosophers have understood the machine in a quite straightforward way, on the basis of human calculation – and the scientist is taken to be an engineer; the machine, a concrete manifestation of theories. The construction of a machine is taken to be a secondary operation; the 'simple application of a knowledge conscious of its import and certain of its effects' (KL 102/76). Canguilhem says that the relationship between technique and science is not as straightforward as this would suggest. He moves through four stages in the lecture (KL 102/76).

The first part of the lecture examines the meaning of the comparison of the organism to a machine. The notion of a machine is 'an artificial construct, a work of man, whose essential function depends on mechanisms'. A mechanism is a 'configuration of solids in motion such that the motion does not abolish the configuration … an assemblage [*assemblage*] of deformable parts, with periodic restoration of the relations between them' (KL 102/76–7). (Canguilhem really does use the French word *assemblage*. Many accounts of Gilles Deleuze and Félix Guattari assume this, though they actually use *agencement*.) Canguilhem goes into more detail about how mechanisms work, but the general point is clear. He cautions that this general principle of machines and their mechanisms is not often found in organisms: 'what is the rule in human industry is the exception in the structure of the organism and in nature' (KL 103/77). So, why does biology insist on using the model of the machine and mechanisms as the way of explaining the organism? For Canguilhem, this is partly because biology does not understand its mechanical models to be structured in quite the same way as this kinematic version. Even those kinematic machines are not self-sufficient, because they require design from someone and some initial impetus, an external 'source of energy', to begin motion (KL 103–4/77–8).

Canguilhem cites Giorgio Baglivi's 1696 text *De praxi medica* at this point, which likens teeth to pliers, blood vessels to hydraulic tubes, the heart to a spring, the lungs to bellows and so on.[17] But what is telling is that Baglivi requires the addition of a machine as a motor, rather than just kinematic mechanisms, in order to operate (KL 104–5/78–9). While the mechanistic model can be traced to Descartes, there are antecedents at least as far back as Aristotle. Aristotle's theory of the movement of animals is quite different from Descartes's work, because Aristotle believed that there was a prime mover – something that sets things in motion. For Aristotle, this is the soul (KL 106/79–80). Descartes, operating at the time of the technological revolution, thinks of mechanism in terms of 'automatons with springs and hydraulic automatons' (KL 106/80).[18] Canguilhem asks 'how do we account for the appearance, so clear and abrupt in Descartes' thought, of a mechanist interpretation of biological phenomena?' (KL 106/80). In part, we need to find this in contemporary economic and political transformations, but this is not straightforward or clear. It is possible that mechanist philosophy is an outcome of these transformations; or that they are together part of a general, global structure.

Canguilhem considers various options. Pierre-Maxime Schuhl provides an analysis based on the history of ideas – in particular, political transformations and the relation between freedom and servitude; Lucien Laberthonnière stresses the relations to Christian theology; and Franz Borkenau a more economically causal relation, through the transformation of artisanal labour to the capitalist model, and the calculative, quantitative model of economics (KL 106–10/80–2).[19] Henrik Grossman shows that this model is historically suspect, because it ignores historical developments which precede modern capitalism: 'Borkenau writes as if Leonardo da Vinci had never existed' (KL 109/82).[20] In summary, Canguilhem is in no sense an economic determinist. His interpretation, based on Grossman's argument, is that mechanization was a development from Renaissance science, and that Descartes rationalized this. He sees this scientific basis as much more plausible than the idea that Descartes implicitly traded on the emergent techniques of a nascent capitalist economy. He stresses that for Descartes mechanics is a *'theory of machines'*, and that the crucial question is making sense of the phenomena of 'spontaneous invention'. His conclusion is that the 'construction of machines' is integrated into his philosophy, rather than that he 'transposed into ideology the social phenomena of capitalist production' (KL 109–10/83; see OC I, 490–8; VR 219–26).

In the next stage of his lecture, he examines the relationship between mechanism and purpose. Two key terms are *finalité* and *finalisme*. Both work with the French sense of *fin*, as end or goal. As such, Canguilhem's translators tend to translate the first term as 'purpose' rather than 'finality', 'to denote the state or quality of nature, machines or organs having a final end or purpose'.[21] Descartes's argument is inseparable from his wider philosophy. His well-known distinction between soul and body, between thought and extendable matter creates a rigid divide, it 'entails affirming the substantial unity of all matter, regardless of its form, and of all thought, regardless of its function' (KL 110/83). Animals cannot judge, speak or invent, and so have no souls, no reason. They have life, but this does not mean they cannot be used instrumentally by humans: 'The mechanization of life, from a theoretical point of view, and the technical utilization of the animal are insepara-ble. Man can make himself master and possessor of nature only if he denies all natural purpose and can consider all of nature, including, apparently, animate nature – except for himself – to be a means' (KL 111/84). Even the human body is a machine, and can be viewed in this way; the human, though, is distinct, because of the soul, the rationality, the *cogito*. Canguilhem's argument is, of course, grounded on Descartes's most famous texts, including the *Discourse on Method* and the *Meditations*, but he also draws on the *Treatise of Man*, 'Description of the Human Body' and some of his correspondence.[22] The full discussion of Descartes's work does not need to be outlined here, but Canguilhem insists on the need for there to be some prior cause, 'a vital original' – either a God or, of course, the soul (KL 113/85). Yet this direction by the soul cannot go beyond the mechanical predisposition of the body, a claim which Canguilhem says 'held sway over the entire theory of automatic and reflex movements until the nineteenth century', though he adds that 'the soul's decision is not a sufficient condition for the movement of the body' (KL 114/86).[23]

Canguilhem contrasts Descartes's position with the critique of vitalism made by Bernard in *Lectures on the Phenomena of Life Common to Animals and Plants*. Bernard denies the idea of a vital force that does work, but concedes that it might act as a director, and suggests that 'the only *vital force* that we could accept would only be a sort of legislative force, but in no way executive'.[24] Canguilhem thus says that Descartes's work does not avoid purpose (*finalité*); mechanism can only explain things if the machines already exist, and it cannot account for their construction. His work is effectively tautological, because 'a tool or a machine is an organ, and organs are tools or machines' (KL 115/87).

Faced with this challenge, the next step is that Canguilhem suggests that we need to turn things around, with a reversal of the traditional relationship between machine and organism. If we are trying to understand an unfamiliar mechanism, then we look to see what it is intended to produce. A sequence of operations, of steps with a logic and purpose, is established to present a particular outcome or goal. It is this function or purpose, rather than the mechanism's form and structure, which will help us to understand it. As such, only after seeing the machine function can we assess it (KL 116/88).

One of the contrasts between a machine and an organism, for Canguilhem, is that, while a machine is constructed, conserved, regulated and repaired from the outside, either by a human or another machine, with organisms we often see 'self-construction, self-conservation, self-regulation, and self-repair' (KL 116/88). With a machine, 'the whole is strictly the sum of the parts', because of its 'rational accounting', 'functional rigidity', 'standardization' and so on. It may seem, therefore, 'that there is more purpose in the machine than in the organism, because the purpose of the machine is rigid, univocal, univalent'. But this is really a narrower purpose, because in organisms there is 'a vicariousness of functions, a polyvalence of organs' (KL 117/88–9). Canguilhem's examples are childhood aphasia, where the development of other parts of the brain ensure language; or the stomach, which was thought to be the organ of digestion but was discovered also to be 'an internal secretion gland'; or an experiment in which a rabbit placenta was grafted into an intestine and came to term (KL 117–18/89; see EHPS 269; VR 126–7). Narrow purpose, therefore, better describes the machine; whereas an organism has 'greater latitude of action' (KL 118/90).

Canguilhem then argues that this links to his concerns with the pathological and monstrosity or deformity (*monstruosités*). 'There is no machine monster ... no mechanical pathology', whereas 'life tolerates monstrosities'. Monsters, whatever else they are, are living beings. Physics and mechanics have no distinction between the normal and the pathological, which is a distinction only for living beings (KL 118/90). These questions are more fully discussed in chapters 2 and 4, but the point is significant. Canguilhem shows how work in experimental embryology challenges the idea that a seed, or an egg, contained a kind of machinery that would inevitably lead to a specific outcome – the growth of an organ or organism. While Descartes held that view, work by a range of biologists has shown this is not the case (KL 119/90–1).

Moving into the last part of the lecture, Canguilhem explores the philosophical consequences of this reversal. He suggests that one of the key things we need to do is to understand the machine, rather than simply explain it. To understand it is to situate it in the context of human history, itself embedded in an understanding of human life, though this needs to be understood not as a merely natural phenomenon (KL 120/91–2). Perhaps surprisingly, Canguilhem then turns to Kant and, in particular, his *Critique of the Power of Judgement*, to show 'the irreducibility of the organism to the machine and, symmetrically, the irreducibility of art to science'. He notes that, while French writers had not looked for a philosophy of techniques in Kant, many German writers since 1870 have done so (KL 121/92). Indeed, he suggests that it is ethnographers, rather than engineers, who have made more significant – if still minor – contributions to our understanding of the construction of machines. Philosophers have tended to be more interested in the philosophy of science, rather than of technique, and it is in ethnographers that we can find attention 'to the relationship between the production of the first tools, the first devices [*dispositifs*] for acting on nature, and organic activity itself' (KL 122/93). The partial exception, for Canguilhem, is Alfred Espinas's work on technology, though he stresses how important Ernst Kapp's work is to Espinas.[25] Kapp and Espinas show, trading on earlier work, that 'the first tools were no more than prolongations of human organs in action', which may explain things like flints, clubs or levers, but struggles to understand the complexities of fire or the wheel (KL 123/94).

Canguilhem also references André Leroi-Gourhan's 1945 book *Milieu et techniques*, which he suggests contains 'what is today the most striking example of a systematic and duly detailed attempt to bring biology and technology together' (KL 124/94).[26] The model proposed is in opposition to the idea that 'technical invention consisted in the application of knowledge', but rather traces antecedent technologies to more modern ones. The locomotive is not so much 'an application of pre-existing theoretical knowledge' but a development of the cylinder and piston model used to drain mines. Further back, the spinning wheel is an ancestor, and this in a biological sense. Canguilhem quotes several passages from Leroi-Gourhan's work which show how what we call 'invention' is often the application of old techniques to new problems, and comments on this: 'One sees how, in the light of these remarks, Science and Technique must be considered not as two types of activity, one of which is grafted onto the other, but where each borrows from the other sometimes its solutions, sometimes its problems. The rationalization of techniques makes one forget the irrational origin of machines' (KL 125/95).

The biological understanding of the organism is therefore helpful in understanding machines, rather than the reverse. For further support, Canguilhem turns to Georges Friedmann's work on industrial mechanization.[27] Technology should be understood as a 'universal biological phenomenon', rather than 'only an intellectual operation of man' (KL 126/96). Instead, therefore, of asking how the organism can be considered as a machine, in either its structure or its functions, it is necessary to ask how that viewpoint emerged (KL 127/96). What was a problem of biology becomes a philosophical, or historical, question. And it is that question that Canguilhem's lecture attempted to answer: 'We have proposed that a mechanist conception of the organism is no less anthropomorphic, in spite of initial appearances, than a teleological conception of the physical world. The solution we have tried to defend has the advantage of showing man in continuity with life through technique, prior to insisting on the rupture for which he assumes responsibility through science' (KL 127/97).

He recognizes that this might be seen to reinforce nostalgia and critiques of technology and progress, but says that he did not intend 'to come to their aid' (KL 127/97). He is insistent that an appropriate approach to life, and the organism, will have to reject the mechanistic approach.

Far bigger questions than he can address are raised by these reflections, and Canguilhem would discuss some in later lectures. In 1961, for example, he discussed the way that models and analogies could be used in the physical sciences.[28] This work was difficult, he said, but 'it seems even more difficult to agree on the role and significance of models in the biological sciences, and even to agree on the definition of such models' (EHPS 305; RAM 507). There are different kinds of models, from mathematical ones to the more common analogical ones (EHPS 305; RAM 507). Of these, the mechanical can be useful in zoology, especially regarding locomotion (EHPS 306–7; RAM 508): 'The Cartesian animal-machine remained a manifesto, a philosophical war-machine: it did not constitute the programme, scheme, or plan of construction of any particular description of function or structure' (EHPS 309; RAM 510).

In 1966, he again returned to the relation between machines and the living, suggesting that the living organism remains irreducible to the machine. One reason is that machines do not assemble themselves, or create other machines. While the development of technology in the last fifty years may blur the point he is making here, he still insists that in an absolute sense the distinction holds. What he calls the 'living machine' should be associated 'with what eighteenth-century thinkers liked to call a *machinist*, an inventor or builder of machines' (EHPS 326; VR 296). That position is confined to the living being. It is a view which, he suggests, brings Kant and Bernard close together. 'A machine, Kant says, is a whole whose parts exist for one another but not by one another. No part is made from any other; in fact, nothing is made of things of the same type [*espèce*] as itself. No machine possesses its own formative energy' (EHPS 326; VR 297). He later comments that this 'recalls Hegel's observation in his *Logic* that it is the whole which creates the relation among its parts, so that without the whole there are no parts' (EHPS 332; VR 301).[29] These kinds of considerations are lost in a merely mechanistic approach.

'The Living and its Milieu'

Canguilhem's third lecture of this series was under the title 'The Living and its Milieu'. Like *milieu*, *le vivant* is not easy to render in English. It can mean simply 'life', but that is the translation of *la vie*, and *le vivant* really refers to something that is living, 'the living', or a 'living being'. However, 'living being' does not have a specifically ontological or existential sense for Canguilhem.[30] Canguilhem begins by suggesting that the concept of milieu, as 'a category of contemporary thought', is crucial to a range of fields – geography, biology, psychology, technology, economic and social history – and is central to understanding 'the experience and existence of living beings'. Nonetheless, its understanding and historical formation is not clear, and thus it falls to philosophy to elucidate the concept (KL 129/98). This cannot be simply a set of comparative descriptions, historically conducted, but a critical analysis which shows 'their common point of departure' and how this has implications for a philosophy of nature (KL 129/98).

The focus is on the period since 1800, and the relation between organism and milieu. 'Milieu' was imported from mechanics to biology, in the late eighteenth century, even if the notion preceded the use of the term. Figures from Isaac Newton and the Encyclopaedists to Lamarck and Buffon, Saint-Hilaire and Comte, Balzac and Hippolyte Taine play a role in this story, from the concept to the plural word, to the singular, to its use in literature, history and related fields. In mechanics, the term relates to what Newton analysed as fluid, or the ether, the medium of action, in which things are in the middle – the second sense of *milieu*. In biology, Canguilhem suggests that the idea came from Lamarck, but its use in the post-1870 neo-Lamarckian biology, as a term both 'universal and abstract', came from Taine's use in history. Taine had suggested that milieu, along with race and moment, were the 'three principles of the analytic explanation of history' (KL 129–30/99).[31]

Canguilhem spends some time tracing the different uses of the concept and the word in these and other writers. The relations are complicated but he suggests various lineages – Newton used the concept in his physics but also in his optics, and this may be its route into biology; Buffon taught Lamarck, and is the link between him and Newton; but Buffon also had links to what Canguilhem calls the anthropo-geographers, who in France were continued by Montesquieu.[32] Buffon's animal ethology draws on both Newtonian mechanics and this anthropo-geography, and Canguilhem wonders if this means that both have some earlier common origin, given the ease with which they were combined in a single theory (KL 131–2/100–1). By the time we reach Comte, who thought the term was his own, there is an attempt to see the interrelation between organism and its environment – 'the total ensemble of exterior circumstances necessary for the existence of each organism'.[33] However, Canguilhem cautions that Comte does not quite reach a dialectical understanding, and 'apart from the human species, he holds the organism's action on the milieu to be negligible'. For Canguilhem, this is due to his continuing to operate within the Newtonian model of 'action and reaction' (KL 133/101–2). 'In sum, the benefit of even a cursory history of the importation of the term milieu into biology during the first years of the nineteenth century is that it accounts for the originally strictly mechanistic acceptance of the term' (KL 134/102).

As Canguilhem discusses in a later essay, Comte plays a fundamental role here. Canguilhem contends that, somewhat surprisingly, Comte's work was more up-to-date with regard to biology than to mathematics or physics, a judgement Canguilhem takes from Paul Tannery (EHPS 61).[34] This was, however, despite Comte having a professional interest in mathematics, and only an amateur interest in biology (EHPS 62). He followed courses on medicine and biology, and he excelled at summarizing their contents (EHPS 62–3; VR 237–8). At the time, biology was being asserted as a separate science, distinct from the physical sciences, though this was in tension in his work. Comte's analysis of the milieu showed that it could not be a separate science; his analysis of the organism indicated that it must be (EHPS 64–5; VR 241). Canguilhem suggests that, although milieu 'was in common use in seventeenth- and eighteenth-century mechanics and the physics of fluids, it was Comte who, by reverting to the word's primary sense, transformed it into a comprehensive, synthetic concept that would prove useful to later biologists and philosophers' (EHPS 65; VR 241). In Comte's view, 'an organism was an *indivisible* structure of *individual* parts. Actual living things were not "individuals" in any simple sense.' This is the basis of his opposition to cell theory (EHPS 66; VR 242).

For Canguilhem, 'Comte claimed to have achieved comprehensive, critical insight into the biology of his time' (EHPS 69; VR 245). What were his contributions? Among them, Canguilhem suggests, was to have 'eliminated a metaphysics of purpose [*finalité*] from biology: following [Georges] Cuvier the principle of "conditions of existence" replaced that dogma of final causes, and the only relation assumed to exist between an organism and its milieu, or between an organ and its functions, was one of compatibility or fitness, implying nothing more than the viability of the living being' (EHPS 69; VR 245). Equally, 'Comte held that all pathological phenomena could be explained by the laws of physiology, by a generalization of the principle borrowed from Broussais. Thus, he argued that the difference between health and disease was a matter of degree rather than kind – hence medicine should base its actions on the analytic laws of anatomophysiology' (EHPS 69–70; VR 245). In Canguilhem's judgement, Comte was able to get beyond a merely cumulative approach to the living being. There was a risk that biology was seen as analogous to chemistry, with the composition of living beings seen as comprised of organic molecules or other building blocks. The vital principle was important beyond this, and Comte was able to discern that life is 'necessarily a property of the whole organism' (EHPS 79; VR 240).

Canguilhem also suggests that Comte sheds important light on social questions: 'human history is the development of human nature, understood as a plurality of virtualities of which the passage to act operates at different speeds ... Life and human experiences are an aspect of the biological co-relation between organisms and their milieux' (EHPS 83). He also claims that 'Comte named physiology the general science [*science*] of organized bodies. Physiology was not just a science [*savoir*] recently instituted for the study of living man, one whose method could serve as a model of the study of living man in society. More than that, the content of physiology was to become the nucleus of a new science [*savoir*]' (VR 248).[35] It follows from this that 'politics is like medicine in that both are disciplines in which perfection requires observation' (VR 248).[36]

In the debate between Lamarck and Darwin, there are tensions around the use of the term (KL 135–8/104–6), but Canguilhem suggests that 'Darwin is more closely related to the geographers, and we know how much he owed to his voyages and explorations. The milieu in which Darwin depicts the life of the living is a bio-geographical milieu' (KL 138/106). This leads Canguilhem to a discussion of nineteenth-century geography, including Carl Ritter's 1817 *Comparative Geography* and Alexander von Humboldt's 1845 *Kosmos*, which he suggests 'stand for the birth of geography as a science conscious of its method and dignity' (KL 138/106).[37] Their work combines, he suggests, the 'science of the human ecumene since Aristotle and Strabo' and 'the science of mathematical geography' (KL 138/106). It would take too much space to trace all the tantalizing lineages which Canguilhem hints at here, but he is interesting on the relation between geography and history in Ritter, and the role of travel and bio-geography – both of plants and animals – in von Humboldt.[38] He suggests that, in both authors, the question of 'the relations between historical man and milieu' is a general study, looking at the world and humanity as a whole, and that this in turn influenced thinkers as diverse as Friedrich Ratzel, Jules Michelet and Taine. In their wake, 'doing history came to consist in reading a map, where the map is the figuration of an ensemble of metrical, geodesic, geological, and climatological data, as well as descriptive bio-geographical data' (KL 139/107). In this light, we might reflect on the hyper-specialism of geography today, and the rigid divide between human and physical geography.

In parallel to this work, there is the study of animal ethology, in which 'the mechanistic explanation of the organism's movements in the milieu succeeded the mechanistic interpretation of the formation of organic forms', a model which has a lineage through to behaviourist psychology (KL 140/107–8). Yet in a wholly mechanistic model, where are the living beings? 'We see individuals, but these are objects; we see gestures, but these are displacements; centres [*centres*], but these are environments [*environnements*]; machinists, but these are machines. The milieu of behaviour coincides with the geographical milieu; the geographical milieu, with the physical milieu' (KL 141/108).

Taken so far, a reversal was inevitable, and Canguilhem finds it in geography, from physical processes to botanico-geography and the interactions of animals and humans with their environments: 'Geography has to do with complexes – complexes of elements whose actions mutually limit each other and in which the effects of causes become causes in turn, modifying the causes that gave rise to them' (KL 141/109). It is especially the case with the human, who, 'as a historical being, becomes the creator of a geographical configuration; he becomes a geographical factor' (KL 142/109). Canguilhem finds an analysis of these factors in French geographers like Albert Demangeon and Vidal de La Blache, and the historian Lucien Febvre of the *Annales* School.[39] In their work, he suggests, there is 'no pure physical milieu', but always one shaped by and, in turn, shaping the human – a position he sees also in Georges Friedmann's work, in animal psychology, and in pragmatists like John Dewey (KL 142–3/109–10). There are developments in *Gestalt-theorie*, notably in Kurt Koffka's 'distinction between the milieu of behaviour and the geographical milieu' (KL 143/110).[40] The nuances of all this work are in danger of being lost, but we must remember this is a lecture, not a fully referenced book.

Canguilhem then turns to the early twentieth-century work of Goldstein and Jakob von Uexküll. They make a reversal of the relation 'with a lucidity that comes from a fully philosophical view of the problem' (KL 143/110). What they show is that a living being can be studied in an experimentally controlled environment, which is to make a milieu for it, while recognizing that the living being also makes that milieu. Canguilhem notes, though, that Goldstein is critical of the fully integrated relation of the living being and its milieu, suggesting that this would mean it was impossible to analyse the relations (KL 144 n. 1 / 177 n. 29).[41] Biologically, there is a similar relation between the parts of an organism, and between the organism and its milieu.

Von Uexküll uses several related German words to describe his work. '*Umwelt* designates the milieu of behaviour [*comportement*] proper to a certain organism; *Umgebung* is the banal geographical environment; *Welt* is the universe of science. The milieu of behaviour proper to the living (*Umwelt*) is an ensemble of excitations, which have the value and signification of signals' (KL 144/111). Von Uexküll is insistent on this point – the excitations need to be noticed, not just to occur, and this requires the living being's interest. In other words, it is the living being that generates the excitation, anticipated and reacted to: 'A living being is not a machine, which responds to excitations with movements, it is a machinist, who responds to signals with operations' (KL 144/111). Even though those reactions might be studied in a physico-chemical way, the biologist is concerned with why, in a plurality of potential excitations, the living being retains and reacts only to some, and how, therefore, 'its life rhythm orders the time of this *Umwelt*, just as it orders space' (KL 145/112). This is a reversal of some earlier figures, who saw those elements as ordering the life.

Von Uexküll's most famous experiment concerned the *Umwelt* of the tick, an insect which lived on the warm blood of mammals.[42] The tick will wait – and in experimental conditions waited for eighteen years – until the right signal: that a mammal is below the tree branch on which they are waiting. They will then drop onto the mammal, seek out an accessible route to the blood, suck the blood and release their eggs to reproduce. The trigger is the smell of rancid butter from the animal's cutaneous glands; experimental conditions can cause this reaction, but if the warmth of the animal and its blood cannot be found, the tick will climb back up and wait again (KL 145–6/112–13).[43] Goldstein critiques the 'mechanical theory of reflexes', showing that the 'reflex is not an isolated or gratuitous reaction', but 'a function of the opening of a sense to stimulations, and of its orientation with regard to them' (KL 146/113). Experimental conditions can create isolated stimuli, but these do not necessarily trigger the same responses:[44] 'The relation between the living and the milieu establishes itself as a debate (*Auseinandersetzung*), to which the living brings its own proper norms of appreciating situations, both dominating the milieu and accommodating itself to it' (KL 146/113; see KL 23/9). But this shows why the experimental situation can only give, at best, a partial piece of information. It can be 'a chapter in physics', but 'in biology, everything is still to be done' (KL 147/113).

In the final stage of this history, Canguilhem turns to near-contemporary debates surrounding heredity and hybridization, developing from the work of Gregor Mendel. In this, there is discussion of 'morphological characteristics', and an 'acceptance of the autonomy of the living in relation to the milieu' (KL 147/114). Canguilhem quotes Albert Brachet's claim that 'the milieu is not, properly speaking, an agent of formation, but rather of realisation', but recognizes that 'the conception of the total autonomy of hereditary genetic material has been criticized' (KL 147–8/114).[45] This discussion of hereditary characteristics inevitably leads to a further discussion of Lysenko. Canguilhem suggests that the polemics around his work – and he seems to imply the arguments both for and against – were 'at least as ideological as scientific' (KL 148/115–16). Indeed, his presentation of Lysenko's work is part positive, recognizing that Lysenko was correct to suggest that it was possible to change the 'hereditary constitution of the organism, wrongly supposed by geneticists to be stable ... Heredity would thus be the assimilation, by the living, over the courses of succeeding generations, of exterior conditions' (KL 148/115). This does not mean that the work was fully Lamarckian, since Lamarck insisted that adaptation was 'the initiative of the organism's needs, efforts, and continual reactions' (KL 149/115). There were certainly challenges to Lysenko's claims, but Canguilhem is more convinced by the Mendelian notion that it is 'the spontaneous character of mutations' that limits attempts at engineering them. He notes that 'the technical – that is, agronomic – aspect of the problem is essential', but that Mendel's work 'tends to dampen [*modérer*] human – and specifically Soviet – ambitions for the total domination of nature and to limit the possibility of intentionally altering living species' (KL 149/115). There is an inevitable politics to a recognition of the 'milieu's determining action', since, if the milieu can be changed, there appears the possibility of a change to human nature, and a progressive change at that. It is for this reason that Mendel could be 'presented as the head of a retrograde, capitalist, and idealist biology' (KL 149/115).

As Canguilhem brings his ambitious lecture to a conclusion, he returns to the relation between biology and mechanics, or the physico-chemical approach. He reinforces that, initially, 'the biological notion of the milieu combined an anthropo-geographical component with a mechanical one' (KL 150/116). Indeed, the anthropo-geographical component was broad enough to encompass the mechanical, since it included astronomy, which had of course been the model for Newton's mechanics: 'What gives meaning to the geographical theory of milieu is the theory of universal sympathy, a vitalist intuition of universal determination' (KL 150/116). Yet, while in debates about this before Copernicus, Kepler and Galileo took the earth as the centre, since them it has been decentred:[46] 'From Galileo and Descartes on, one had to choose between two theories of milieu, that is, between two theories of space: a centred, qualified space, where the mi-*lieu* is a centre; or a decentred, homogeneous space, where the *mi*-lieu is an intermediary field' (KL 150/117).

The struggle between these theories can be found in many writers, and Canguilhem suggests Pascal perhaps exemplifies this. Canguilhem encourages us to think of the living being in the centre of its milieu, but recognizes that these are plural: 'the milieu on which the organism depends is structured, organized, by the organism itself ... within what appears to man as a single milieu, various living beings carve out their specific and singular milieux in incomparable ways' (KL 152/118). This is true for the human too, with the milieu as the world of perception, experience and action. The human experience of a milieu and engagements with it are specific, particular, to them (KL 152/118).

However, at the same time, the human as scientist (*savant*) 'constructs a universe of phenomena and laws that he holds to be an absolute universe' (KL 152–3/119). In Einstein, we find the exemplary, ideal representation of the fundamental laws of this universe. It is a human conceit to make us prefer our 'own milieu over the milieux of other living beings, as having more reality and not just a different value' (KL 153/119). But the woodlouse or the grey mouse has just as much reality, and ideally the term 'real' should only be applied to the absolute universe, not to the specific milieu of the human. If we are true to the scientific project, then this must encompass the human as well as all other living beings. In this, we must avoid reducing living beings to mechanical, physical and chemical processes, and keep sight of the indissociable living being (KL 153/119).

Canguilhem concludes with some reflections on what this means for the relation of science itself to human life. If 'innate knowledge is rightly refused, the birth, becoming, and progress of science must be understood as a sort of enterprise as adventurous as life' (KL 153/119). So, rather than any idea that 'reality contains the science of reality beforehand, as part of itself', which would be absurd, Canguilhem suggests:

But if science is the work of a humanity rooted in life before being enlightened by knowledge, if science is a fact in the world at the same time as it is a vision of the world, then it maintains a permanent and obligatory relation with perception. And thus the milieu proper to men is not situated within the universal milieu as contents in a container. A centre does not resolve into its environment. A living being is not reducible to a crossroads of influences. From this stems the insufficiency of any biology that, in complete submission to the spirit of the physico-chemical sciences, would seek to eliminate all consideration of sense from its domain. From the biological and psychological point of view, a sense is an appreciation of values in relation to a need. And for the one who experiences and lives it, a need is an irreducible, and thereby absolute, system of reference.

(KL 154/120)

The importance of these lectures is fourfold. First, they show how his concerns with questions of medicine, explored in his formal training and his thesis *The Normal and the Pathological*, were situated within a wider interest in biology. Second, that his concern with biology was to understand it in a philosophical register. Third, that his understanding of the relation between the organism and its milieu was entirely opposed to a reductive, mechanistic approach. The relation between biology and other sciences is important, but cannot be reduced to them, for they fail to grasp the irreducible nature of life. It is for this reason that he insists on the importance of vitalism, even if he recognizes some of the problems of this approach. There are problems of method and in application, notably in politics, on both left and right. Finally, the key theme of the milieu is discussed in detail, a topic to which Canguilhem would return many times. All of these questions are approached historically.

4

Physiology and the Reflex

Knowledge of Life was first published in 1952, and brought together a number of texts by Canguilhem – the 'Cell Theory' essay from 1945, the three lectures from 1947 discussed in chapter 3, 'Experimentation in Animal Biology' from 1951, and 'The Normal and the Pathological' from 1951. To these essays, Canguilhem added a new Introduction. The book was republished in 1962, with the addition of a lecture from that year, 'Monstrosity and the Monstrous'. Although the book does work as a single volume, for thematic and chronological reasons the texts have been disaggregated in the discussions here. This chapter and the next two discuss all these works, along with *La formation du concept de réflexe* and some related works from the 1950s and 1960s.

This chapter will first discuss the Introduction to *Knowledge of Life*, followed by the 'Experimentation in Animal Biology' lecture from 1951 as a way into Canguilhem's views on physiology. This is followed by a discussion of his most sustained historical study, *La formation du concept de réflexe* and some related essays. Chapter 5 then discusses his work on regulation and his work on psychology and cognition. Regulation is a theme that continues his interest in physiology, but also bridges his concerns with biology and medicine, the animal and human, and whether the life of an organism can be used to understand society. Chapter 6 looks at his collaborative work on evolution and the question of monstrosity, concluding with his two 1966 lectures titled 'The New Knowledge of Life', published in *Études d'histoire*, which return to some themes from the book *Knowledge of Life*.

As Marrati and Meyers note in their useful foreword to the English translation, *Knowledge of Life* is a 'subtly problematic title of his book', because 'we are confronted not with a single history but with many singular histories'.[1] The difficulties also inhere in the word 'knowledge'. In the opening lines of the book, Canguilhem claims that 'to know is to analyse' (KL 9/xvii). Here, he uses the French *connaître*, but elsewhere in this book he frequently uses the other main French verb *savoir*. Canguilhem seems to use the two terms interchangeably, along with their substantives *connaissance* and *savoir*, unlike Foucault, for whom there is a strict and crucial distinction. *Savoir*, for Canguilhem can also mean 'science', though he uses the French *science* too. What then, for Canguilhem, is the relation between knowledge and life, a fundamental issue for a study of the life sciences?

Canguilhem argues that 'we accept far too easily that there exists a fundamental conflict between knowledge [*connaissance*] and life, such that their reciprocal aversion can lead only to the destruction of life by knowledge or to the derision of knowledge by life' (KL 9/xvii). Canguilhem rejects the choice that he thinks this view leads to – 'between a crystalline (i.e. transparent and inert) intellectualism and a foggy (at once active and muddled) mysticism' (KL 9/xvii). Instead, he suggests, 'the conflict is not between thought and life in man, but between man and the world in the human consciousness of life' (KL 9–10/xvii). It is worth noting that, though problematic to our modern ears, like many of his generation, Canguilhem uses man, *l'homme*, to stand for the human race as a whole; and *humain* as the corresponding adjective.

Knowledge then, is a process of trying to make sense of the relation between the human and the world. It disentangles, interrogates and doubts. It 'consists in the search for security via the reduction of obstacles; it consists in the construction of theories that proceed by assimilation' (KL 10/xviii). This sense of knowledge takes its cue from the goal or purpose of knowledge, which is to make sense of the human/world interaction, to undo the experience of life, to make sense of its failures and abstract from it a set of lessons and strategies. Knowledge 'is thus a general method for the direct and indirect resolution of tensions between man and milieu' (KL 10/xviii). The question of knowledge is peculiar to humans, but there is a problem in claiming that non-human animals are blind because they do not see like us, or stupid because they do not act in the same way we do. 'Doubtless, the animal cannot resolve all the problems we present to it, but this is because these problems are ours and not its own. Could a man make a nest better than a bird, a web better than a spider?' (KL 10/xviii).

Knowledge is, in a sense, a tool – and a tool developed for a specific purpose related to life and its challenges. Canguilhem draws again on Leroi-Gourhan's work, suggesting that tools are never 'created wholly for a use yet to be found, on materials yet to be discovered'.[2] Science, then, like religion and art, is 'specifically human'; all are 'ruptures with simple life', and yet none should be seen as depreciations of life (KL 10–11/xviii, see 124–5/94):

> Even if knowledge is the daughter of human fear (astonishment, anxiety, etc.), it would not be very insightful to convert this fear into an irreducible aversion to the condition of beings that experience fear in the crises they must overcome so long as they live. If knowledge is the daughter of fear, this is for the domination and organization of human experience, for the freedom of life.
>
> (KL 11/xix)

This, therefore, explores the relation between the two terms of the book's title: 'Thus, the universal relation of human knowledge to living organization reveals itself through the relation of knowledge to human life. Life is the formation of forms; knowledge is the analysis of in-formed matter' (KL 11/xix).

The knowledge of life is, of course, most closely associated with the discipline of biology, though that term emerges quite late in the histories that Canguilhem is tracing (EHPS 215). In *Les mots et les choses*, known in English as *The Order of Things*, Foucault would trace in detail the shift from natural history to biology as an analysis of life, along with parallel developments in the understanding of labour and language.[3] Canguilhem notes that Darwin, for example, called himself a naturalist all his life, and Thomas Huxley was 'inventor and popularizer of the very word biology'.[4] What is the relation of biology to the other sciences? Canguilhem discusses this through some reflections on the thought of Darwin's contemporary, Bernard: 'In these propositions we find the wavering that is habitual in Bernard's thought: on the one hand, he senses the inadequacy of analytical thought to any biological object; on the other, he remains fascinated by the prestige of the physico-chemical sciences, which he hoped biology would come to resemble, believing it would thus better ensure the success of medicine' (KL 12/xx). The model of the physical sciences is a recurrent theme, though its inadequacy is also crucial. Rationalism must recognize its own limits, and when it looks at life draw on that experience:

'Intelligence can apply itself to life only if it recognizes the originality of life. The thought of the living must take from the living the idea of the living.' Canguilhem uses a passage from Goldstein's study of the organism to illustrate this point: 'it is evident that no matter how much [the biologist] employs the analytical method for obtaining real knowledge, for real insight into the depths of nature the departure from the "immediately given" will always dominate'. He comments on this to suggest that 'to do mathematics, it would suffice that we be angels. But to do biology, even with the aid of intelligence, we sometimes need to feel like stupid beasts [*bêtes*] ourselves' (KL 12–13/xx).[5]

Claude Bernard and experimentation in animal biology

For the 1951 lecture titled 'Experimentation in Animal Biology', it is worth knowing that the French word *expérience* signifies both 'experiment' and 'experience' in English. This creates difficulties in always knowing which sense is intended. The two meanings can lead to quite different understandings of science, and to a French critique of experimentation in Bachelard, Koyré and Canguilhem.[6] Canguilhem begins this essay by again stressing the importance of Bernard.[7] Indeed, he says that Bernard's *Introduction to the Study of Experimental Medicine* (1865) is often seen, following Bergson, as the equivalent to Descartes's *Discourse on Method*. What Descartes's text is for the material sciences, Bernard's study is for the life sciences. Yet Canguilhem stresses that both these works are read as instruction manuals, 'without making the effort to reinsert either of these works into the history of biology or mathematics'. Equally, such readings do not undertake the work of interrogating the relation between a text which is written by a 'gentleman scientist addressing other gentlemen' and the actual practice of specialists in the field (KL 17/3; see EHPS 155). As he suggests elsewhere, to commemorate Bernard's *Introduction* after 100 years, for a 'historian of biological sciences or a historian of medicine, is to interrogate the sense and direction [*sens*] of our enterprise' (EHPS 127). This is in the sense of what it is to commemorate the anniversary of a scientist's work, which scientists are chosen, and the national focus, but also the crucial nature of this specific work (EHPS 127–8). Such an approach represents much of Canguilhem's work practice. He is asking how we should understand texts in their context, which may be detached from the practice that they describe; and how do we situate texts within the history of that knowledge and practice.

Canguilhem's analysis of Bernard is important in multiple ways. Canguilhem suggests that, before Bernard, the choice for a biologist was either to see biology as physics, or to make a sharp distinction:

> The Newton of the living organism was Claude Bernard, in the sense that it was he who realised that the condition of possibility of the experimental science of living beings was not the scientist, but the living being itself, which provided the key to deciphering their own structures and functions. Rejecting both mechanism and vitalism, Bernard was able to develop techniques of biological experimentation suited to the specific nature of the object of study.
>
> (EHPS 149; VR 267)

Two striking things about Bernard's work are that the first part of his *Introduction* seems only to restate things that we feel we have always known, and that much of the work on it reduces it just to that first part on laboratories, rather than the second and third parts which relate to experimentation in biology. This is another point Canguilhem attributes to Bergson (KL 17–18/3–4; see also EHPS 163–71). If we take the example of muscular contraction, we can see what Bernard intended, but also challenge the idea that he was an innovator. Many medical manuals attribute the codification of experimental operations to Bernard, or even the invention of these procedures, even if he was explicit in showing how he was building on previous work (KL 19/5). Canguilhem underscores, though, that Bernard shows us that 'biological functions can only be discovered through experimentation', and he stresses that Bernard's book on experimental physiology in medicine is more explicit on this point (KL 20/6).[8]

Bernard, therefore, saw himself as the 'founder of experimental medicine' (EHPS 138; VR 279), a term which is closer to the English notion of experimental physiology. He challenged people who clung to old ontology or vitalism, showing that the issues they raised as objections could also be explained by this new approach (EHPS 139; VR 279). Although he stressed his own innovation, he saw Francis Bacon as something of an inspiration. Bernard suggested that 'there were great experimentalists before all precepts of experimentalism',[9] which Canguilhem sees as a self-comment too, and suggests that Bernard broke with the rules of mathematical physics in his route to science (EHPS 144–5; VR 262). Canguilhem suggests that Bernard's work is a compromise between two conflicting models of biology – that of Bichat and that of Magendie (VR 271).[10] Magendie was Bernard's master, and one of the forerunners of experimental medicine – though there may be even earlier figures (EHPS 134). But their styles were different: 'Magendie had asserted, refuted, condemned – for him, life [*le vivant*] was a mechanical phenomenon and vitalism an aberration' (EHPS 139; VR 280).

Bernard was not, therefore, a vitalist, mechanist or materialist (VR 273).[11] He was 'a determinist without being a mechanist', and sought to understand 'vitalism as an error rather than a folly' (EHPS 139; VR 280). Bernard's view was that 'Medicine is the art of healing, but it must become the science of healing. The *art* of healing is empiricism. The *science* of healing is rationalism.'[12] He wanted to base his work in physiology on 'assumptions and principles stemming from the domain of physiology itself, from the living organism'. This was crucial to the foundation of this science on its own terms, rather than taking forward ideas from other sciences such as physics and chemistry. While those other sciences were crucial, it was essential to base physiology on the living organism (EHPS 145; VR 263).

One example was the role of the liver. Bernard suggested that the fact that its role in synthesizing glucose had not been discovered before was because scientists were not looking in the right place. He said its importance and easy observation had not been identified because it needed to be seen as a functioning organ within an organism, not abstractly through the model of other sciences (EHPS 145–6; VR 263).[13] Hence his claim that 'Neither anatomy nor chemistry can answer a question of physiology. What is crucial is experimentation on animals, which makes it possible to observe the mechanics of a function in a living being, thus leading to the discovery of phenomena that could not have been predicted, which cannot be studied in any other way.'[14]

Canguilhem also discusses how Bernard defended vivisection as mode of access to the internal milieu (EHPS 147–8; VR 268). Bernard's approach challenged people like Cuvier, Comte and Blainville for whom 'comparative anatomy was a substitute for experimentation, which they held to be impossible because the analytic search for the simple phenomenon inevitably, or so they believed, distorts the essence of the organism, which functions as a whole' (EHPS 150; VR 270). In contrast, Bernard 'saw comparative anatomy as a prerequisite for developing a general physiology on the basis of experiments in comparative physiology. Comparative anatomy taught physiologists that nature laid the groundwork for physiology by producing a variety of structures for analysis. Paradoxically, it was the increasing individuation of organisms in the animal series that made the analytical study of functions possible' (EHPS 150–1; VR 270–1). As Bernard recognized, complex animals have differentiated organs and other physiological features. Indeed, Canguilhem suggests that the more complex the organism, the more separate the 'physiological phenomena' are. The morphologically complex organism means that functions are distinct, and can be more easily studied. It is harder to study these functions in an elementary organism, where 'everything is confused because everything is confounded'. While 'the laws of Cartesian mechanics are best studied in simple machines', in contrast, 'the laws of Bernardian physiology are best studied in complex organisms' (EHPS 151; VR 271).

Canguilhem further notes that Bernard made a significant distinction between '*laws*, which are general and applicable to all things ... and *forms* or *processes*, which are specific to organisms. This specificity is sometimes termed "morphological", sometimes "evolutionary"' (EHPS 158–9; VR 273). This 'organic guiding principle', Canguilhem suggests, may well have been the key to Bernard's philosophy of biology.

> That may be why it remained somewhat vague, masked by the very terms it used to express the idea of organization – vital idea, vital design, phenomenal sense, directed order, arrangement, ordering, vital preordering, plan, blueprint, and formation, among others. Is it too audacious to suggest that with these concepts, equivalent in Bernard's mind, he intuitively senses what we might nowadays call the antirandom character of life – antirandom in the sense not of indeterminate but of negative entropy?
>
> (EHPS 159; VR 274)

Canguilhem even suggests that Bernard almost anticipated modern developments in biology, in what is called 'genetic code', following the influence of cybernetics. Canguilhem notes that 'the word "code", after all, has multiple meanings, and when Bernard wrote that the vital force has legislative powers, his metaphor may have been a harbinger of things to come'. Yet he was still a man of the nineteenth century, and 'glimpsed only a part of the future, for he does not seem to have guessed that even information (or, to use his term, legislation) requires a certain quantity of energy'. Canguilhem adds that, while 'he called his doctrine "physical vitalism", it is legitimate to ask whether, given his notion of physical force and his failure to grant the "vital idea" the status of a force, he really went beyond the metaphysical vitalism that he condemned in Bichat' (EHPS 160; VR 275).[15]

More broadly, experimentation raises a thorny question of the specific nature of organisms. The questions which Comte, Bernard and Canguilhem are thinking about include whether one individual animal can stand for a species, and whether something we learn about one species tells us about another. With specificity, this raises the problem of generalization, and Canguilhem suggests that, instead of simply giving the name of a person to a law or phenomenon, we should also add 'the name of the animal used for the experiment' (KL 26/11). He makes this point explicit: 'What is important here is that no experimentally acquired fact (whether it deals with structures, functions, or comportments) can be generalized either from one variety to another within a single species, or from one species to another, or from animal to man without express reservations' (KL 27/12). These reservations include the generalization of the facts 'from one variety to another … from one species to another … from animal to man' (KL 27/12). That is clear between species, but what about within them? Canguilhem addresses this question of individualization with similar doubt: 'how could one be certain in advance that two individual organisms are identical in all aspects when, although they belong to the same species, each has a unique combination of hereditary characteristics owing to the conditions of their birth (sexuality, fertilization, amphimixis)?' (KL 28/13). (Amphimixis is the process of sperm and egg cell fusion). There are ways around these challenges, but often these depend on processes which are themselves questionable theoretically.

These two challenges – of whether one individual animal can be generalized to a species, and whether one species can reveal something about another – are quite straightforward to grasp, but Canguilhem also adds there are issues for experimentation concerning the totality of the organism and 'the irreversibility of vital phenomena' (KL 26/11). Totality is important for at least two reasons – the question of isolation, and that of integrity:

> Is it possible to analyse what determines a phenomenon by isolating it, given that we are operating on a whole, which, as such, is altered by any attempted removal? It is not certain that, after the ablation of an organ (ovary, stomach, kidney), an organism is still the same organism minus one organ. On the contrary, there is reason to believe that we are dealing with a very different organism, which we cannot easily 'superimpose', even partly, onto the control organism.
>
> (KL 29/13)

This is to do with the polyvalent nature of organs, and that 'all phenomena are integrated' (KL 29/13–14).[16] Irreversibility is also significant: 'If the totality of the organism is a difficulty for analysis, then the irreversibility of biological phenomena, either from the viewpoint of the being's development or from the viewpoint of the adult being's functions, constitutes a further difficulty for chronological extrapolation and prediction' (KL 30/14). He adds that 'over the course of its life, an organism evolves irreversibly in such a way that the majority of its presumed components are – if one considers them separately – full of potentialities that do not reveal themselves at all under normal conditions of existence' (KL 30/14).

So, experiment cannot be used to prove things across species, within species, with an organism, and 'Bernard remarked that if no animal is absolutely comparable to another of the same species, neither is the same animal comparable to itself if examined at different moments in its life' (KL 30–1/15). But this last principle is not without potentially productive results. Canguilhem notes that immunity was part proved after Louis Pasteur injected a chicken with an old cholera culture, and then with a fresh one (KL 31/15). But even this work needs to know and appreciate its limits, restressing his key point about the irreducible nature of biology: 'Today, one would have to be quite uninformed of the methodological tendencies of biologists – even those biologists least inclined to mysticism – to believe that anyone can honestly boast of having discovered, by physico-chemical methods, anything more than the physico-chemical content of phenomena, whose biological meaning escapes all techniques of reduction' (KL 32/16).

What, then are appropriate techniques for biological experimentation? Canguilhem suggests some general principles, whether they concern modifications to the milieu of an organism, or on an embryo in development (KL 33/17). Nonetheless, these still raise questions of how we can generalize from an experimental, artificial procedure. Equally, the observation, by itself, influences the thing being observed. Most fundamentally, 'how do we conclude the normal from the experimental?' (KL 34/18). There are even more significant questions when it comes to these issues with humans. Biology is in a privileged position here, among other specialized studies of humanity – offering the potential for not just knowledge, but also transformation. However, there are, of course, ethical norms regarding experimentation on humans, and biology has a specific responsibility in this regard. While some of its scientific approaches are shared with other sciences such as physics and chemistry, biology also affects 'the identity of man as both subject of knowledge and object of action', and so it has to have a philosophical sense of what humanity is – effectively, a philosophy. Canguilhem describes this as 'philanthropic impulses combining with humanist hesitancies' (KL 35/19).

In conclusion Canguilhem suggests that:

> The problem of human experimentation is no longer a simple problem of technique but a problem of value. As soon as biology concerns man no longer merely as a problem but as an instrument for research into solutions concerning him, it becomes a matter of deciding whether the benefit of knowledge is such that the subject of knowledge could consent to becoming the object of its own knowledge. We have no difficulty recognizing here the always-open debate regarding man as a means or an end, as an object or a person. This is to say that human biology does not contain within itself the answer to questions concerning its nature and meaning [*signification*].
>
> (KL 38/21)

In his discussions of physiology, particularly the role of experimentation in it, Canguilhem has raised some important issues around innovation and precursors, generalization and comparison, and the balance between theory and technique. This inquiry also continues his interrogation of the relation between vitalism and mechanism in biology, and the relation of biology to other sciences. These themes also come to the fore in his detailed study of the reflex.

The formation of the concept of reflex

La formation du concept de réflexe was Canguilhem's primary doctoral thesis in philosophy, submitted in 1955 and supervised by Bachelard.[17] It is his most sustained historical examination of a specific scientific question, but raises significant philosophical and biological issues as well. It returns to a major theme in *Knowledge of Life*, which was submitted as the secondary thesis alongside this work: the stake of biology as a mechanistic or vitalist science. Canguilhem is concerned with showing that the organism should not be understood as a machine, and that it is not straightforwardly conditioned by the milieu of which it is a part. The specific analysis is of the reflex function: of how a part of the body of an organism reacts to some stimulus without conscious thought. There are many kinds of reflexes, from pulling a hand or foot away from a hot object, to sneezing, scratching or muscular ones such as the knee-jerk or patellar reflex. The term 'reflex arc' is used to describe how a stimulus to a nerve is carried through neurons to the spinal cord and this affects the muscles, even before the message reaches the brain. The action is the reflex itself, the arc the neural pathway that produces it or makes it possible.

The nineteenth-century tradition in biology considered that it was Descartes who had invented the concept of the reflex, and that it was part of his mechanistic analysis of the body. Canguilhem instead explores the present theory of the reflex to examine its roots, and shows that its first uses were actually elsewhere and of a different register. His analysis demonstrates that it was actually Thomas Willis who developed a theory which was taken up and elaborated as the dominant stream of the tradition. This is not just a question of priority in terms of ideas, though Willis (1621–75) and Descartes (1596–1650) were near-contemporaries. Rather, seeing Willis as the key figure is significant because Willis was proposing a vitalist, rather than mechanist, approach to the question (FCR 171–2; see KL 93/68).[18] Accordingly, this work is not just a historical study of the notion of the reflex, but, perhaps more importantly, a critique of the classical idea that the organism is mechanistically conditioned by its milieu.[19] As Braunstein puts it, 'the mechanistic conception of the milieu was not only unfair [*injuste*], it was false'.[20]

Part of the problem is that existing histories have failed to differentiate a range of different aspects of the question of the reflex. Canguilhem suggests that 'descriptions of automatic neuromuscular responses, experimental study of anatomical structures and their functional interactions, and the formulation of the concept and its generalization in the form of a theory' have been muddled together (FCR 3; VR 179). Nonetheless, he is clear that his purpose is not to 'right wrongs, like some scholarly zealot', but to draw out some more general conclusions which may be of use for 'the history of science and for epistemology' (FCR 3; VR 179).

He outlines two prejudices which he thinks are widespread but which have powerful effects. The first is that 'people are disposed to believe that a concept can originate only within the framework of a theory or at least a heuristic inspiration, homogeneous to that in terms of which the observed facts will later be interpreted' (FCR 3; VR 179). The second is more specific to biology, but is grounded on the belief that 'the only theories that have led to fruitful applications and positive advances in knowledge [*acquisitions positives*] have been mechanistic in style' (FCR 3; VR 180). As a result, in the second half of the nineteenth century, 'the reflex was universally regarded by physiologists as the element of composition of all animal movement, according to laws of complexity which a mechanistic conception of life robbed of all teleological value' (FCR 3).

The mechanist approach dominated not just the present, but the way that the present conceived its own past, thinking that this 'mechanical explanation of animal life could have been discovered and studied only by a mechanist biologist. If the logic of history pointed to a mechanist, physiology provided the name: Descartes' (FCR 4; VR 180). Yet there are multiple problems. Canguilhem questions whether the 'logic confirmed the history or the history inspired the logic', and says that there is a retrospective approach being taken here. Descartes discussed involuntary movement, and did so in a mechanistic way. Some of what he analysed would, later, be called 'reflexes'. Yet this does not mean, Canguilhem underscores, that 'Descartes had described, named and formulated the concept of the reflex'. Rather, 'the general theory of the reflex was elaborated in order to explain the class of phenomena that he had explained in his own fashion' (FCR 4–5; VR 180).

The point of Canguilhem's argument is not to take away from Descartes – though he thinks that his reputation can probably survive this specific case. His point is that a reading 'more attentive to the truth than to the glory of Descartes' is appropriate (FCR 5). In the history of sciences, he thinks that 'rights of the logic of history must not be allowed to obscure the rights of logic'. We need to understand theories in the ways that 'contemporaries interpreted the concepts of which those theories were composed' (FCR 5; VR 180). We need to investigate 'conceptual filiations' in a different way:

> Rather than ask who was the author of a theory of involuntary move-
> ment that prefigured the nineteenth-century theory of the reflex, we
> ask what a theory of muscular movement and nerve action must
> incorporate in order for a notion like reflex movement, involving
> as it does a comparison between a biological phenomenon and an
> optical one [i.e. reflection, from which 'reflex' is derived], to have a
> sense of truth.
>
> (FCR 5–6; VR 180)

This approach, employed by Canguilhem in the study, does not *discover* Thomas Willis – for Canguilhem recognizes that he was mentioned by some studies of the question before – but rather 'confirms his legitimate right to a title that had previously been open to doubt or challenge' (FCR 6; VR 181–2).

The point of this study is therefore not just to tell, and correct, the history of the reflex. It has a more metahistorical point too. One of the chapters is a discussion of the 'History of the History of the Reflex', looking at the way that the story came to be established in the nineteenth and twentieth centuries (FCR 132–67). Already in the book's foreword, Canguilhem had talked of the research he had done for the book, initially in Strasbourg and then in Paris – the latter 'more slowly and less easily'. A large part of his focus is on the secondary and historical literature, rather than just the primary sources from the seventeenth and eighteenth centuries (FCR 1). He singles out the importance of Franklin Fearing's 1930 book *Reflex Action: A Study in the History of Physiological Psychology*, though notes that this was unavailable in France and out of print in the United States, so eventually he had to make use of a microfilm (FCR 1–2). (The book has since been reissued.[21]) He notes that the work arrived late in his research, and so he was able to use it to 'confirm or contradict our own positions'. He praises the work as 'incontestably' the best and most complete study of the subject (FCR 37).

But the key difference between their work, he suggests, is that, while Fearing provides a very broad and comprehensive survey, he is interested in a much more specific analysis of the seventeenth and eighteenth centuries (FCR 2). In particular, he wants to challenge Fearing's unexamined claim of Descartes's importance (FCR 141). In the study of the history, Canguilhem does acknowledge the valuable work that has gone before, but is equally critical of other works, such as Charles Scott Sherrington's *The Integrative Action of the Nervous System*, which provides no history, never mentions Willis, discusses Descartes only twice – and even that regarding different issues (FCR 151).[22] Some other works are assessed more generously, of course, including a work by Hoff and Kellaway on 'the early history of the reflex' which is described as 'in our eyes, a model of the genre' (FCR 154),[23] and he similarly appreciates the work of J. F. Fulton on reflex control of movement and the nervous system (FCR 45, 51 n. 1, 151).[24] The Hoff and Kellaway article, in particular, closely parallels some of Canguilhem's argument.

The conventional story was that Descartes discovered the reflex action, and that he was 'the first to publish a systematic study of the phenomena of involuntary movement' (FCR 37). In this work, he associated these movements with notions we would today call reflexes. Yet, as Canguilhem shows, to make the assumption that this means he had that notion himself is based on the suggestion that, though Descartes did not name it as such, this 'concept is implicit' in his work (FCR 37).[25] Another study is clearer on this point: 'the notion of reflex action is traceable to Descartes, but the term hardly so. The term is traced more clearly to Thomas Willis. Descartes did not consider all our acts to be "reflex".'[26] Yet Canguilhem patiently explains that Descartes's analysis of the phenomena of involuntary movement is filled with conceptualizations which would today be rejected, and his work contains 'neither the term nor the concept of reflex' (FCR 52; VR 184). Descartes did not make a distinction between sensory nerves and motor nerves, sees pushing as creating a push, and pressing a press. He therefore saw involuntary movement as 'different from action in all of its elements and phases' (FCR 35; VR 183). As such, the Cartesian approach is 'certainly mechanical, but it is not the theory of the reflex'. The reflex theory requires 'homogeneity between the incident movement and the reflected movement', but in Descartes 'the opposite is true: the excitation of the senses and the contraction of the muscles are not at all similar movements with respect to either the nature of thing moved or the mode of motion' (FCR 41; VR 184).

Yet the challenge to Descartes is not just on this point: it is also a challenge to mechanism much more generally. Canguilhem concludes the chapter on Descartes by suggesting that the issues at stake go beyond merely the analysis of the reflex, and concern biology more generally: 'We will say that only a metaphysician can formulate the principles of a mechanistic biology without risk of immediate contradiction, even though that contradiction will ultimately be discovered. Historians of biology have seldom noticed this, and even fewer biologists. It is more regrettable, however, when they are philosophers' (FCR 56; VR 236).

If Descartes cannot resolve the question of this concept, Canguilhem turns instead to Willis. He asks why he is ordinarily neglected by historians of biology and medicine in this account? Part of the reason, Canguilhem contends, is that they are so focused on the Cartesian lineage that they cannot imagine that Willis's ideas are comparable (FCR 57–8). There are differences in multiple registers. These extend to their way of thinking about mechanics – Willis did work in this register, but this does not mean he was a mechanist. Descartes inherited the 'statics of Archimedes and the dynamic of Galileo', even with modifications. In contrast, Willis is more concerned with anticipating energetics, and while Galileo's and Descartes's notion of the dynamic was a cannonball, Willis's notion was 'the blast of the powder' (FCR 60). Canguilhem traces the way that Willis took the analysis of the heart and the circulation of the blood from Harvey, and this meant that he also saw the heart simply as a muscle, neither privileged nor pre-eminent. Willis distinguished between circulation of the blood and its fermentation: 'a mechanical phenomenon and a chemical phenomenon' (FCR 60–1; VR 185–6). The fidelity to Harvey, and his 'more chemical than mechanical notion of animal spirits' (FCR 65; VR 187), however, important though it was, is not sufficient to mark the distinction between Willis and Descartes. This can be found in his rejection of the differentiation of sensory and motor functions of the nerves, and in his imagination in working through some of these different conceits. Willis understood the nerves as radiation, a ray of light emanating from the 'flame of the blood'. Just like light, 'the nervous discharge was instantaneous', and even the 'final stage of transmission, the excitation of the muscle by the nerve, supported the comparison'. The contraction of the muscle was caused by a 'spasmodic intramuscular explosion', the light becomes fire, and 'in this physiology the nerves are not strings or conduits but fuses' (FCR 65–6; VR 187–8).

Willis is truly the first to formulate 'the concept of reflex movement', and this can be found in his 1670 text 'De motu musculari' (FCR 67–8).[27] Canguilhem stresses the crucial point: 'We are truly here in the presence of a concept, because we find its definition – a definition at once nominal and real.' There are various stages here: 'In addition to the object being defined, we have a defining proposition, which fixes its meaning. We have a word that establishes the adequacy of the defining proposition to the object defined.' In the particular instance of the scratch, we have both

> the thing, in the form of an original observation, a cutaneous reflect of the cerebrospinal system, the scratch reflex; the word, reflex, which has become classic even if improperly as an adjective and a noun; and the notion, that is, the possibility of a judgment, initially in the form of an identification or classification and subsequently in the form of a principle of empirical interpretation.

There are nuances of his analysis here which might be addressed, but he suggests that, 'in sum, concerning the reflex, we find in Willis the thing, the word and the notion [*la chose, le mot, et la notion*]' (FCR 68–9; VR 188–9).

Now, there are over 100 pages of his study beyond this point, and Canguilhem's analysis does not end with the emergence of this concept in this initial, somewhat inchoate, form. His point is to displace Descartes and restore Willis's importance. Willis indicates the future, rather than providing all the answers. His research also focuses on different mammals and birds, whereas Descartes's work tended to look at animals in general, in a less differentiated way (FCR 73–4). Willis is a determinist, but his focus is chemical in terms of 'nerve and muscle functions'; he begins to work out the relation between the cerebrum and cerebellum and sensory-motor functions; and he reorganizes the relation between the body's peripheries and its centre. While Descartes saw the central motor as commanding peripheral action, Willis saw the peripheral as the manifestation itself (FCR 77–8). In summary, for Canguilhem, 'if Descartes made, as we have seen, some steps on the path which led to the concept of the reflex, Willis was able to make all the steps himself' (FCR 78).

In the remainder of the book, Canguilhem discusses later theorizations and additions at some length. There are discussions of internal heat and, in particular, experiments on decapitated animals, which helped to demonstrate the importance of the spinal column. One chapter is on Johann August Unzer and Georg Prochaska, whom Canguilhem thinks are crucial to the overall story, as are Robert Whytt, Julien Jean César Legallois and Jean Astruc.[28] He summarizes the development that all this work takes us to at the beginning of the nineteenth century:

> Thus, by 1800 the definition of the reflex concept was in place, a definition ideal when considered as a whole but historical in each of its parts. It can be summarised as follows (with the names of the author who first formulated or incorporated certain basic notions indicated in parentheses): a reflex movement (Willis) is one whose immediate cause is an antecedent sensation (Willis), the effect of which is determined by physical laws (Willis, Astruc, Unzer, Prochaska) – in relation to the instincts (Whytt, Prochaska) – through reflection (Willis, Astruc, Unzer, Prochaska) in the spinal cord (Whytt, Prochaska, Legallois), with or without concomitant consciousness (Prochaska).
> (FCR 131; VR 194)

The importance of Canguilhem's summary analysis here is not to stress the individual names as much as to show how the lineage of concepts – this relation between thing, word and notion – is complicated and multi-faceted. In the conclusion, he stresses that all the names listed as crucial to the story he tells are 'biologists and doctors of the animist or vitalist tendency, with the sole exception of Astruc' (FCR 169). Descartes studied neuromuscular functions through mechanics, Willis through chemistry, yet contemporary work on the physiology of the nervous system was trying to reconcile their insights (FCR 170). Returning to his earlier claim about

Descartes, Canguilhem suggests that his status as 'the greatest French philosopher' is safe from this study of his limitations in physiology. Equally, Willis cannot lay claim to having a work equivalent to the *Geometry* or the *Meditations*, and Canguilhem is not seeking to find it (FCR 170–1).

Towards the end of the book he returns to Bachelard, and sees this study as a contribution to understanding what his supervisor called 'the contemporary past [*passé actuel*]'. This work is history in the sense not of what Bachelard called a 'palaeontology of a lost scientific spirit', with the view of resurrecting these abandoned ideas, but of '*a recurrent history*, a history which is illuminated by the *finality of the present*' (FCR 167).[29] As he approvingly cites Viktor von Weizsäcker: 'The reflex is a concept with a development, but also one which we still use today. So the historical study intervenes in current research on nature' (FCR 167).[30]

In the references and bibliography, some of Canguilhem's breadth of interest can be seen. There are works of medicine, biology, neurology, psychology and philosophy, referenced in English, French, German and Latin. The appendix of the book includes bilingual passages from works by Willis in Latin and Canguilhem's French translation (FCR 173–92). His philosophical references are not just to Descartes and other classical figures, but continue into the twentieth century. While you might expect references to the work of his supervisor, as well as referencing some of Bachelard's philosophical works (FCR 201), Canguilhem also uses his books *The Psychoanalysis of Fire* (FCR 197) and *Earth and Reveries of Repose* (FCR 198).[31] Merleau-Ponty's *The Structure of Behaviour* was also important, alongside Goldstein's *The Organism* (FCR 164, 200–1). In 1977, for the re-edition of the text, Canguilhem added a page of more recent references on the same topic (FCR 202). What is striking here is that, while Canguilhem's reading is wide and in several languages, few of the new English sources make any reference to his 1955 study.[32] Some continue to make the claims for precedence and lineage which Canguilhem's study patiently exposes as erroneous.[33]

A decade after the publication of *La formation du concept de réflexe*, Canguilhem returned to the topic to write a piece on the concept in the nineteenth century.[34] There is little sense of a self-critique in this work: it is more of a continuation. While chapter VII of the book had discussed the historiography of the nineteenth century, here he is concerned with extending the analysis of the book in terms of the developments in the concept and the understanding of it in that century. This is not a strict divide – Sherrington, who was criticized for his ahistorical approach in the book (FCR 151), is discussed both there and here as a scientist (EHPS 302–3; VR 200–1). Canguilhem argues that 'the nineteenth century did not *invent* the concept of the reflex, but it *corrected* it' (EHPS 295). This correction was not done logically, but through experimentation (EHPS 295–6). Canguilhem returns to the definition quoted above, where he had set out the state of things in 1800 (FCR 130–1; VR 194). Now he suggests that 'we can see precisely what elements stood in need of correction' (EHPS 296; VR 195).

The purely technical aspects of this reading need not detain us here. But the essay is striking for two reasons. First, for its closing lines, which suggest that it was only in the nineteenth century that the concept of the reflex was 'purged of any teleological implications', and that it was no longer seen 'as nothing more than a simple mechanical reaction'. As he concludes, 'through a series of corrections, it had become an authentically physiological concept' (EHPS 304; VR 201–2). The second reason is its definitional labour, important for his wider historical work.

> In speaking of a 'concept', we understand, according to use, a denomination (*motus reflexus, reflexion*) and a definition, in other words a name with a meaning [*sens*], capable of filling a function of discrimination in the interpretation of certain observations or experiences relative to movements of organisms in a normal or pathological state. In the genus of movement, the concept of the reflex delimits a certain species.
>
> (EHPS 295)

This relation between denomination and definition is the one familiar from the German tradition of *Begriffsgeschichte*, conceptual history, developed by Reinhart Koselleck. Koselleck labels the relation, which works both ways between words and concepts, as semasiological and onamasiological: which concepts are implied by words, which is the question of sense or meaning; and what words are used to denote specific concepts, which is the question of designation.[35] Crucially though, in a way that the German tradition is less effective at, Canguilhem brings this relation between word and concept to bear on a third term, the practice. In the passage cited above, Canguilhem stresses the relation between thing, word and notion, the tripartite elements of the concept (FCR 68–9; VR 188–9). As such, while *Begriffsgeschichte* looks at the relation between word and concept, Canguilhem sees the concept as containing the word alongside the object and the notion.

Physiology as science

Canguilhem remained interested in the relation of biology to the sciences throughout his career. He stressed that experimentation could provide insights, but was also interested in the question of explanation, which required the generalization of observed phenomena. Canguilhem argued that the use of models was less common in the biological sciences than the physical sciences, and that analogical models were more common than mathematical ones.

> Experimentation is analytic and proceeds by discriminating among the determining conditions by varying them, all other factors being supposed unchanged. The model method allows the comparison of entities which resist analysis. Now, in biology, analysis is less a partition than a liberation of entities of a smaller scale than the initial one. In this science, the use of models can legitimately pass as being more 'natural' than elsewhere.
>
> (EHPS 311; RAM 513)

There are further contrasts between different parts of the natural sciences. It is, for one, harder to resist the temptation for a model to be taken as a representation in biology than in physics (EHPS 313; RAM 514). Canguilhem adds that 'one cannot, at the present time at least, speak of a mathematical biology in the sense in which, as we have seen, one has for a long time spoken of mathematical physics' (EHPS 314; RAM 515). This does not mean that there is no relation between biology and mathematics – there was arithmetic and geometric biology in antiquity; statistical biology more recently, but not an algebraic biology (EHPS 314; RAM 516). Indeed, Canguilhem sees this as one of the significant developments of his time: 'the formation of a mathematical biology, which does not necessarily mean an analytic biology, but a biology in which non-quantitative concepts, like those of topology for instance, permit not only the description of phenomena, but the theorization of them' (EHPS 317; RAM 519). As he discusses in some 1966 lectures, for this understanding we need 'a non-metric theory of space, a science of order, a topology. To understand the living being at the scale we place them at, one needs a nonnumerical calculus [*calcul*], a combinatorics, a statistical machinery [*calcul*]' (EHPS 362–3; VR 317). This means that Aristotle's suggestion that mathematics did not help with biology has been proved untrue, but only because of the developments of geometry, calculus and other branches of mathematics. (EHPS 363; VR 318).

In 1963, Canguilhem wrote 'The Constitution of Physiology as a Science' as an Introduction to a collection on physiology (EHPS 226–73). He begins with Jean Fernel's 1542 description of physiology as 'the constitution of man for as long as he enjoys favourable health'.[36] Canguilhem comments: 'It scarcely matters here that Fernel's idea of human nature is more metaphysical than positive. The point to be noted is the birth of physiology in 1542 as a study distinct from, and prior to, pathology, which itself was prior to the arts of prognosis, hygiene and therapeutics' (EHPS 226; VR 91). In the 1960s, Canguilhem suggests that the current meaning of physiology is 'the science of the functions and functional constants of living organisms' (EHPS 226; VR 91). Canguilhem's purpose is to trace the transition between these moments.

He is particularly interested in the relation between anatomy and physiology, noting that, in the eighteenth century, physiology was described by Albrecht von Haller as anatomy in motion. Canguilhem contends that it is more complicated: 'anatomy is the description of organs, while physiology is the explanation of their functions' (EHPS 227). In this period, we also see comparative anatomy eclipsing simple anatomy (EHPS 228). Regarding the eighteenth to nineteenth centuries, he notes that 'if the physical–chemical sciences exerted growing influence on research in physiology, it was mainly because physiologists found the techniques of those sciences indispensable as research tools, though not necessarily as theoretical models' (EHPS 231; VR 104).

Again, Canguilhem turns to the role of experimentation in physiology, and he suggests that Bernard's idea that this was when it became a science is important, even if not taken literally (EHPS 231; VR 104). Yet this was not the only aspect that was important, especially if the instrumental aspect was the focus: 'Some historical overviews and methodological manifestoes give the impression that instruments and the techniques that used them were somehow *ideas*. To be sure, using an instrument obliges the user to subscribe to a hypothesis about the function under study' (EHPS 232; VR 107):

> I cannot agree with those historians of physiology, professional as well as amateur, who would outdo even Claude Bernard's open hostility to theory by ascribing all progress in nineteenth-century physiology to experimentation. The theories that Bernard condemned were systems such as animism and vitalism, that is, doctrines that answer questions by incorporating them.
>
> (EHPS 232–3; VR 108)

Canguilhem adds that his work on medical experimentation was 'a long plea on behalf of the value of ideas in research, with the understanding, of course, that in science an idea is a guide, not a straitjacket [*idée fixe*]' (EHPS 233; VR 108).

Canguilhem gives a number of examples, which are 'drawn from various fields of research, [which] show that experimentalists need not pretend to be pure empiricists, working without ideas of any kind, in order to make progress. The acquisition of scientific knowledge requires a certain kind of lucidity' (EHPS 235; VR 110). He quotes Bernard at this point: 'the experimentalist who does not know what he is looking for will not understand what he finds'.[37] Physiology uses such a number of methods, drawn from a range of other sciences, that it cannot be defined by the specificity of its method, but rather by the problem which it approached through these methods. However, he notes that any contemporary physiologist would accept the definition given by Bernard or Max Verworn: that physiology is 'the explanation of life' (EHPS 238). This is in part because there are understood to be multiple modes of life for animals, and the scale of biological organism is a crucial aspect of the definition of physiology (EHPS 238–9). In his pithy summation: organisms 'are mechanisms capable of reproducing themselves' (EHPS 261; VR 117). Physiology is at the heart of his concerns, not least for the reason that it is a contribution to the knowledge of life.

5

Regulation and Psychology

The problem of regulation

The question of regulation in Canguilhem straddles the boundary between his work on biology and on medicine, between the animal and the human. It was a theme he returned to several times over the years. In 1955 he contributed a piece to the *Cahiers de l'Alliance israélite universelle*, based on a talk with a subsequent discussion (OC IV, 643–72).[1] Canguilhem asks his audience to reflect on the nature of the 'relations between the life of the organism and the life of a society'. Is the comparison or assimilation (*assimilation*) of a society to an organism 'anything more than a metaphor', or does it imply 'a substantial kinship?' (WM 102/67). In part, the lecture is a critique of the cybernetic work of Norbert Weiner, whom Geroulanos and Meyers describe as its 'quiet target'.[2]

He immediately points out that the relation works both ways: 'This permanent comparison of a society to an organism derives from a temptation that is, in general, doubled by the inverse temptation – that of comparing the organism to society' (WM 102/68). Canguilhem gives various examples. One is the way that disease or pathological disorder could be understood as sedition in Greek thought. For Alcmaeon of Croton, 'to explain the nature of the disease in the organism, he brought over a concept of sociological and political origin' (WM 102–3/68).[3] Another example would be the way that the division of labour, understood as a social question – whether seen in a positive or negative light – also found its way into biological thought: 'physiologists found it natural to speak of the division of labour concerning the cells, the organs, and the devices [*appareils*] that make up a living body' (WM 103/68). Other examples include Bernard's talk of the 'social life' of cells, or Ernst Haeckel's 'cell State' or 'Republic of cells' (WM 103/68).[4]

There are thus multiple examples of the 'movement from sociology to biology', and 'there has always been an exchange of ambiguous figures of speech between sociology and biology' (WM 104/68). Only a historical examination can sometimes shed light on precedence of the terms, when they are used by both areas. Examples might include 'crisis' or 'constitution', both of which are used in each field. With 'crisis', Canguilhem is clear: 'it is a concept of medical origin – it is the concept of a change that, signalled by certain symptoms, intervenes in the course of an illness and that will indeed decide the life of the patient' (WM 104/69). But with 'constitution', used in both fields, 'the term has always been ambiguous, equivocal' (WM 104/69). The point of showing this relation is to demonstrate that 'organicism', one particular way of comparing society to an organism, is not the only way that the terms have been related. This theory was 'short-lived', and was an explicit formulation of the relation, but there were questions of this relation much further back (WM 105/109).

But exploring the simple relation between the social and the biological is not the principal aim of this lecture. Through it, we are able to gain insight into the question of the 'structure and functioning of a society', and, especially, 'the reforms to be carried out once the society in question is affected by grave disorders' (WM 106/69) – 'the idea of social medication, the idea of social therapeutics, the idea of remedies for social ills' (WM 106/70). But 'organic and social disorders' and the remedies employed are 'radically different when an organism is concerned and when a society is concerned' (WM 106/70). This applies to the relation between these phenomena in their sick, diseased state and the ideal one. With an organism, we know what the ideal is: 'the ideal of a sick organism is a healthy organism of the same species'. So, while there might be debate about the nature of the sickness, there is agreement about the goal: 'the restoration of the organism to its healthy state' (WM 107/70). This is distinct when we think of societies, and 'their disorders and unrests', and how we might deal with 'ills and reforms'. The difference, in part, is because 'what we debate is how to know its ideal state or norm' (WM 108/70).

We know the 'purpose [*finalité*] of an organism'; but the purpose of society is always something up for debate. The ideal state of a society is in question, not fixed, and we are likely to find quicker agreement on the nature of problems than the solution to them. For biology, 'one could say that in the organic order, the use of an organ, a device, an organism is patent'. In contrast to this, in society, what is sometimes – or even often – obscure is the nature of a problem. 'From the social point of view, it seems on the contrary that abuse, disorder, and the evil [*mal*] are clearer than normal circumstances' (WM 108–9/70–1); 'One could say that in the social order, madness is more clearly discerned than reason, whereas in the organic order, it is health that is more easily discerned and determined than the nature of the illness' (WM 109/71).

Only at this point does Canguilhem turn to the notion of 'regulation' itself. He explains that it is 'a scientist's word, though not really, insofar as everyone knows what a regulator in an old locomotive is and everyone knows what a regulating station is. The concept of regulation is a concept that I would not call familiar, yet not forbidding, either' (WM 110/71). Of course, with sixty years elapsed since Canguilhem's lecture, these are hardly familiar things to us today. As his translators note, the locomotive regulator is like a throttle – it controls the steam flow to the pistons driving the wheels. A 'regulating station' is 'a military command centre established to control all movements of personnel and supplies into or out of a given area'.[5] As Canguilhem suggests, the life of an organism 'is made possible by the existence in the organism of a set of apparatuses [*dispositifs*] or mechanisms of regulation whose effect consists precisely in the maintenance of this integrity, in the persistence of the organism as a whole' (WM 110/72).

Regulation is a fairly recent concept, finding a formulation in Bernard's physiology, although Hippocratic medicine also anticipated some of these ideas. The notion that the organism had some means of self-medication or compensation is old, and modern physiology has provided more explicit formulations and compensations: 'An organism comprises, by the sole fact that it is an organism, a system of mechanisms of correction and compensation for the divergences and injuries to which it is subjected by the world in which it lives – by its milieu, a milieu vis-à-vis which these mechanisms of regulation allow the organism to lead a relatively independent existence' (WM 111/72). While cold-blooded animals (poikilotherms) are 'slaves to the temperature of the milieu', warm-blooded animals, or homeotherms, have 'a system of regulation that allows them to compensate for differences [between its temperature and that of the milieu], to maintain a constant temperature independent of the milieu's prompts' (WM 111/72).

> The innate moderation of an organism is what Walter B. Cannon called 'homeostasis' – a process of conserving Bernard's notion of an 'internal milieu' (WM 112/72).[6] Indeed, the internal milieu of living beings is always in direct relation with the normal or pathological vital manifestations of organic units. In proportion as we ascend the scale of living beings, the organism grows more complex, the organic units become more delicate and require a more perfected internal milieu. The circulating liquids, the blood serum and the intra-organic fluids all constitute the internal milieu.[7]

This notion of 'internal milieu' is important, Canguilhem suggests, but it is not just that Bernard showed that it exists, but that it is produced by the organism, and that in the higher animals there are two key mechanisms for its regulation – the nervous and endocrine systems. Endocrine systems are internal secretions by the glands. 'The regulations that interested Bernard were physiological regulations', including respiration, secretion, thermoregulation, nourishment and so on (WM 112–13/73).

Bernard suggested that 'the blood is made for the organs. That much is true. But it cannot be repeated too often that it is also made by the organs.' Canguilhem comments that 'what allowed Bernard to propose this radical revision of haematology was the concept of internal secretions, which he had formulated two years earlier'. The blood has a different relation to the lungs than the liver – in the former, the body reacts to the external milieu; in the latter, it reacts within itself (EHPS 148; VR 269): 'we should insist on this point: the *concept* of the internal milieu is the theoretical underpinning of the *technique* of physiological experimentation' (EHPS 148; VR 268). He stresses this claim: 'without the notion of internal secretions, there could be no idea of an internal milieu, and without the idea of an internal milieu, there could be no autonomous science of physiology' (EHPS 148; VR 269; see also VR 272).[8] It was a revolution in biological epistemology, more than in biology.

Cannon's book *The Wisdom of the Body* has an epilogue titled 'Relations between Biological and Social Homeostasis', in which Canguilhem suggests he cannot resist the temptation 'to import into sociology this magnificent concept of regulation and homeostasis, whose mechanism he has described in the course of the preceding pages' (WM 116–17/74). This is the shortest but still the 'weakest part of his book', 'because the majority of comparisons are founded on commonplaces of politics and sociology, whose foundation he does not seek' (WM 117/75). There may be some relation between Cannon's interest in these questions and those of Bergson in *The Two Sources of Morality and Religion* (WM 116–19/74–6).[9]

Ultimately, for Canguilhem, society must not be seen as an organism:

> Concerning society, we must address a confusion that consists in the confounding of organization and organism. That fact that a society is organized – and there is no society without a minimum of organization – does not mean that it is organic; I would gladly say that organization at the level of society is of the order not of organic organization, but of design [*agencement*]. What defines the organism is precisely that its purpose, in the form of its totality, is present to it and to all its parts … a society has no proper purpose; a society is a means; a society is more on the order of a machine or of a tool than on the order of an organism.
>
> (WM 120/76)

Even the fact that a society is 'a collectivity of living beings' does not mean that society is an organism, because 'this collectivity is neither an individual nor a species'. This is politically significant: 'Consequently, not being an organism, society presupposes and even calls for regulations; there is no society without regulation, and there is no society without rules, yet in society, there is no self-regulation. There, regulation is always, if I may say so, something added on and always precarious' (WM 121/77). In summary, he suggests that 'disorder and crisis', rather than 'order and harmony', characterize 'the state of society', understood as a machine or tool: 'It is a tool that is always out of order, because it is deprived of its specific apparatus of self-regulation' (WM 122/77). Even the system of justice has to come from outside: 'justice cannot be a social institution … it is not a regulation inherent in society but a different thing altogether' (WM 122/77).

He breaks off the lecture in a somewhat disordered way, noting that, even if he has 'not succeeded in proving to you that society is not an organism (and besides, in these matters there is no proof)', he hopes he has at least raised some of the dangers, and that 'one must not allow it to resemble an organism, [and] that we must be vigilant toward all these comparisons [*assimilations*] whose consequences you can guess' (WM 124–5/78). If Canguilhem's politics are surely aiming in the right direction – and we might pause to reflect on the audience he is addressing – the demonstration is, as he suggests, somewhat lacking. The more persuasive part of the argument seems to be that concerning the question of regulation in the organism, rather than in society, or the relation of the organism and society. And it is this question of regulation in the organism which is the principal theme of his later work on this topic.

Almost twenty years after this lecture, in 1973, Canguilhem wrote the entry on 'Regulation (Epistemology)' for the *Encyclopædia Universalis*.[10] His entry, for such a general audience, summarizes many of the themes he had treated in more detail elsewhere. He suggests that there are three key elements to the notion of regulation: the interaction of unstable elements; criterion or reference; and comparator. The concept of regulation has a lineage in mechanics, where regulation concerns adjustment to a norm, standard or rule, and derives from a 'regulator' – a term from economy, politics or engineering. In the nineteenth century, that mechanical sense was the predominant meaning, but Antoine Lavoisier had used the term '*régulateur*' at the end of the eighteenth century to think about animal physiology in his experiments on respiration.[11] The mechanical notion moved from the machine to the animal-machine to the human-machine. Various aspects of the body were understood to regulate balance and equilibrium, and this regulation and compensation was concerned with processes such as respiration, temperature, adaptation to milieu, healing and feeding. In that broad understanding, 'regulation is the *biological fact* par excellence'.[12] Today, the term is more broadly used in the natural and social sciences, though especially in biology and economics. Thomas Malthus, for example, used the term in his *Essay on the Principle of Population*, to theorize the relation between population and subsistence resources, but it can also be found in the work of Marx or Comte, looking at the interaction of economic and social relations, or even in Claude Lévi-Strauss's work on modern societies.

Canguilhem continued this work in the essay 'The Development of the Concept of Biological Regulation in the Eighteenth and Nineteenth Centuries' from 1974, first presented to a seminar at the Collège de France.[13] It was republished in *Ideology and Rationality* in 1977, but there was 'a strongly political' version of this work in 1973 delivered as the final lecture of a 'three-lecture course at the Catholic University of Louvain titled "Fin des normes ou crise des regulations?" [The end of norms or a crisis of regulation?]'.[14] The first two lectures were on the question of the '*sauvage*', which means wild as much as savage, and the relation of the abnormal and antinormal. Canguilhem had taught a course on the notion of the 'savage', drawing on anthropologists including Ruth Benedict and Lévi-Strauss, in 1959.[15] The final lecture was entitled 'Regulation as a Reality and as a Fiction' and was critical of both traditional medicine and anti-medicine, as well as discussing antipsychiatry and antipedagogy.[16] David Cooper and Ivan Illich are two of the people with whom this work was in dialogue.

Yet in contrast to this more political reading, the published version of 'The Development of the Concept of Biological Regulation' returns to Canguilhem's more standard approach of the history of thought. He begins with a discussion of Driesch's *Die organischen Regulationem*, published in 1901, and its discussion of the constitution of eggs (IR 81/81).[17] The work was a development of themes in physiology, showing that 'there exist organic functions whose purpose is to control other functions and thus, by regulating certain invariants, to enable the organism to comport itself as a whole' (IR 81/82). These are 'regulatory' functions, and Canguilhem suggests that 'this choice of name culminated a long and difficult history of conceptual change, a history that is not easily recounted' (IR 81–2/82). In contrast to 'cybernetics', a term used in the nineteenth century in relation to the science of government, but which did not get taken up in biology for more than a century, 'regulation' is a term with a much richer history. Beginning with Émile Littré's French dictionary and the *Oxford English Dictionary*, Canguilhem shows that the term 'regulation' and the notion of the 'regulator' are closely related, and that the dominant meanings concern the political and mechanical (IR 82/82): 'Thus we cannot undertake the history of "regulation" without understanding the history of "regulators", which will take us into questions of theology, astronomy, technology, medicine, and even sociology at its inception. It is a history in which Newton and Leibniz are no less implicated than Watt and Lavoisier, Malthus and Auguste Comte" (IR 83/83).

The essay that follows is a *tour de force* regarding just these figures. It begins with the dispute between Leibniz and the Newtonians, of which the famous Leibniz – Samuel Clarke correspondence is but one element. In that dispute, the role of God in the running of the universe was in question: did God, having created the world, sit back and watch, or did he play an active role in surveying, correcting and governing its operation? Regulation by the system, or regulation by the creator – Leibniz and Newton differ fundamentally (IR 83–5/83–5):

Leibniz held that the relation between rule and regulation [*règle et réglement*] (in the sense of the policing of a state [*police de l'État*] or regulating [*réglage*] a machine) is from the beginning a static and pacific relation. There is no disparity between rule and regularity. Regularity is not obtained as an effect of regularization; it is not a triumph over instability or a recovery after degradation. Rather, it is an inherent property. A rule is a rule and always remains so; its regulatory function remains latent.

(IR 85/86)

This view was dominant for some time, and this notion of 'conservation and equilibrium' was validated through experiments in the eighteenth century. Canguilhem uses Koyré's work *Études newtoniennes* to support this claim (IR 85–6/86).[18] The 'Leibnizian idea of regulation, understood as the conservation of initial constants' was dominant, not just in mechanics, but as part of the 'paradigm' for many other sciences, from political economy to physiology. Canguilhem suggests that it 'lay behind what Michel Foucault has called the "enunciative regularities" of an era' (IR 86/87).[19] It took some time before biologists began 'to think of organic regulation in terms of adaptation and not just in terms of conservation or restoration of a closed system' (IR 87–8/87).

The next part of the essay turns to 'Economy, Technology, and Physiology'. Canguilhem notes various intriguing relations – the notion of 'animal economy', despite being seen by some as effectively equivalent to 'animal machine' or 'animal factory', brought in the idea that the animal was in some sense regulated or coordinated in its parts and functions. The term was then turned back to human economy, with the idea of the division of labour. It was not just that organic beings were seen as machines, but that machines were made to be more organic – that is, controlled and regulated from within – and that this in turn influenced physiology, and in particular the introduction of the term 'regulator'. From this complex interrelation of technological advances and reformulations in biology, the notion of regulation developed. Yet, for some time, the Leibnizian notion of a closed system was important, and notions of 'inherent power of restitution or reintegration' or 'powers of self-preservation' dominated the eighteenth century. This, Canguilhem claims, 'led to the emergence of an antimechanist, naturist dogmatism that seriously impeded experimental research in physiology' (IR 89/89). There are complicated and sometimes paradoxical relations here – Canguilhem notes that it was Stahl's work that shaped the

88   Canguilhem

dogmatism, and that it was dependent on Leibniz's mechanics, even though Leibniz and Stahl disagreed on animism and physiology. Crucial, though, in the dispute between them, is Leibniz's equation of the animal body with 'a hydraulic-pneumatic heat engine [*machinam hydraulico-pneumatico-pyriam*]' (IR 89/90).[20]

Canguilhem suggests that this development beyond the relatively simple mechanism of a watch or clock shows a more complex system at work. It was picked up by Lavoisier, who compared 'this engine's properties with respect to maintenance, conservation, and reparation with those of an apparatus [*dispositif*] of equilibrium or mechanical regulation'. His work, along with the papers he wrote with Armand Séguin, are 'the first scientific treatises on regulation' (IR 90/90). François Jacob discusses these themes in *The Logic of the Living* – another good example of Canguilhem drawing on the work of a figure whose writings he had influenced.[21] The three main regulators of an animal are respiration, perspiration and digestion – and these three regulators 'govern' the animal machine: they help to restore equilibrium in the system. Lavoisier was ahead of his contemporaries, but still very much a 'man of his century', and his understanding of regulation was very much within an internal system of nature. While it hints at the relation of the organism and the environment as being aleatory, essentially his view is conservative, maintaining the immutable physical order (IR 90/91).

Some of these principles cross between biology and sociology, and Canguilhem provides details of the shared conceptions between, for example, Linnaeus and Malthus (IR 91–2/92–3). There is a more complicated relation between Comte and Bernard. Comte is described as 'a man of the eighteenth century living in the nineteenth century' (IR 93/94): a figure who drew on work and developed themes in mechanics, medicine, astronomy, biology and sociology. Two key principles dominate Comte's work – 'the exterior regulates the interior; it is the stability of the solar system which, through the mediation of milieux, stabilizes living systems' and 'human history is nothing but the development of a seed [*germe*], the realization of human nature. Progress is nothing but the development of order' (IR 93/94–5). Canguilhem suggests that, for Comte, 'a living system is a system open to the outside world, upon which it depends for nourishment for its so-called vegetative functions and for the information required by the animal functions that in one way or another serve the vegetative' (IR 95/96).[22] As Comte himself put it: 'the milieu therefore constitutes the principal regulator of the organism'.[23]

Bernard claimed that he was 'the first to distinguish between an internal milieu and an external milieu'.[24] Canguilhem agrees that Bernard 'was the first to attach positive content to the concept of physiological regulation' but, nonetheless, he 'used the terms *regulator* and *regulation* quite sparingly, and only, it seems, in connection with the circulation of the blood and the phenomenon of "calorification"' (IR 96/98). Canguilhem spends some time discussing the other words that Bernard used, and the quite specific sense that he gives to *regulator* (IR 96–7/98):

> Such regulation by the interior was quite different from regulation in Comte's sense. For Comte, regulation extended to the organism the benefits of a stabilized and stabilizing exterior. For Bernard, regulation referred to a mechanism for stabilizing internal conditions within limits necessary for the life of the cells,enabling the organism to confront the hazards [*aléas*] of the milieu by compensating for deviations [*écarts*].
>
> (IR 97/99)[25]

The final part of this text discusses some nineteenth-century developments in Germany, but notes that 'after Bernard the term *regulation* entered the vocabulary of physiology. When a word appears in the title of a book or paper, it has been recognized as more than a mere metaphor by the competent scientific community' (IR 99/100). As well as being an indication of this specific concept, that is a general point which is worth considering. The essay ends with the turn of the twentieth century, and Driesch's book:

> By now, the range of biological functions to which the concept of regulation was applied had broadened. One could speak of 'regulations' in the plural, and it was the plural that occurred in Driesch's title. When a word is plural, it indicates the extension of the concept, and some temporary consensus had been achieved as to its meaning. 'Regulation', having begun as a purely mechanical concept, had become a biological concept as well. Later, through the mediation of the concept of homeostasis, it would become a concept of cybernetics.
>
> (IR 99/101)

There are recurrent themes here in the relation between mechanism and biology, and the relation of the individual organism to its milieu. Such themes, albeit in a quite different register, dominate some of his other work on the human sciences.

'What is Psychology?'

In December 1956, Canguilhem gave a lecture at the Sorbonne on the topic of 'What is Psychology?' It was published in 1958 and reprinted in *Études d'histoire*. It was not the first time that he had discussed psychology, which was subjected to a critique as early as his 1930s work with Planet, which had taken issue with the Comtean reduction of the moral sciences to the natural sciences.[26] Canguilhem's challenge to the determinism in his own work on biology, especially the idea that the milieu entirely conditions behaviour, can be seen as a continuation of that work, and his challenge is both physiological and psychological.[27] Braunstein has even suggested that the critique of psychology in the way Canguilhem understood it was because he saw some of its claims about conditioning by a milieu amounting to the same thinking as led to collaboration in the Second World War.[28]

He begins this lecture with the suggestion that psychologists do not often ask what their subject matter is, compared to philosophy: 'For philosophy is much more constituted by the question of its meaning [*sens*] and its essence than defined by an answer to that question ... a reason for humility and not a cause for humiliation' (EHPS 365; WP 37). Do psychologists today – sixty years on from this lecture – have a more self-reflexive attitude?

> The characteristic unity of the concept of a science has traditionally been taken as deriving from the object of that science: the object has been thought of as itself dictating the method to be used in the study of its properties. But in the last analysis this amounted to limiting science to the study of a given, to the exploration of a domain. When it became apparent that every science more or less gives itself its given, and thereby appropriates what is called its domain, the concept of a science began to place more emphasis on method than on object. Or, to be more precise, the expression 'the object of the science' acquired a new meaning. The object of a science is no longer simply the specific field in which problems are to be resolved and obstacles removed, it is also the intentions and ambitions of the subject of the science, the specific project that informs a theoretical consciousness.
>
> (EHPS 366; WP 38)

One response would be a unitary domain – the kind of response given by Lagache in 1947. Canguilhem describes this possible 'unity of psychology' as being defined as 'a general theory of conduct, a synthesis of experimental psychology, clinical psychology, psychoanalysis, social psychology, and ethnology' (EHPS 366; WP 38). But, ultimately, he thinks that this was just a truce between natural, experimental psychology and humanist, clinical psychology, with the emphasis on the second, and animal psychology subsumed into the first (EHPS 366–7; WP 38). Not being a psychologist, Canguilhem takes a different approach, suggesting that a brief history of psychology is necessary in order to understand how we have reached the present moment. He notes that this is narrowed to look at its 'founding orientations', and considered through its connections to 'the history of philosophy and the sciences'. He cautions that it is 'a necessarily teleological history, since its goal is to bring to bear on the question under discussion the presumed original sense of the various disciplines, methods or programmes whose present heterogeneity makes the question a legitimate one to pose' (EHPS 367–8; WP 39).

It is telling that, once again, Canguilhem resorts to a historical analysis to address a contemporary issue. He notes that there was no independent psychology in antiquity, and that work in that area was divided between metaphysics, logic and physics, of which biology is a part (EHPS 368; WP 39). This meant that, for a long time, psychology did not get analysed separately, and only with the seventeenth-century eclipse of Aristotelian physics did psychology become a science of subjectivity (EHPS 369; WP 40). Mechanist physicists are thus crucial to the birth of modern psychology, which emerges as a 'psycho-physics', based on its calculus and relating to its perception (EHPS 369–70; WP 40–1). Descartes and Malebranche are the key figures, but the term 'psychology', as 'science of the self', comes from Wolff in the eighteenth century (EHPS 370–1; WP 41–2).

Yet the work on psychology was not just conducted in works of philosophy – it was, of course, practically driven as well. Canguilhem gives Philippe Pinel credit for founding mental medicine 'as an independent discipline, building it up from its beginnings in the therapeutic isolation of the insane at Bicêtre and La Salpêtrière' (EHPS 374–5; WP 44–5). Pinel was, famously, a figure who removed the chains from the mad at the Bicêtre hospital, and La Salpêtrière was, among other things, where Jean-Martin Charcot conducted his work with hysterics.[29] The term 'psychoanalysis' was coined by Freud in 1896, but it was not just the word for an existing practice. From this moment, Canguilhem suggests, 'psychology is no longer only the science of the intimate, it is the science of the depths of the soul' (EHPS 375; WP 45).

A fundamental issue in the nineteenth century is the constitution of a psychology which is simultaneously conceived 'as nervous and mental pathology, as the physics of the external senses and as the science of the internal and intimate senses, of a biology of human behaviour' (EHPS 376; WP 46). Canguilhem links these changes to his wider work and social conditions. He sees the developments in psychology as linked to the transition in biology to a study of organism and milieu; the end of discrete human order; industrial change and the industry of human being; egalitarianism and the end of social privilege (EHPS 376; WP 46). He provides a brief discussion of Nietzsche's *On the Genealogy of Morality* (EHPS 377–8; WP 46–7), which discusses some of these themes in a summary and indicative way. However, perhaps the most interesting part of Canguilhem's analysis comes in his critique of behaviourism. This has been described as a decision 'to treat someone like an insect', which he notes is a phrase the novelist Stendhal borrows from the naturalist Cuvier (EHPS 379; WP 48).[30] But he turns it back on the discipline too: 'And what if we were to treat the psychologists like an insect, if we were to apply Stendhal's suggestion to the bleak and insipid Kinsey, for example?' (EHPS 379–80; WP 48). Alfred Kinsey, the founder of the Institute for Sex Research at Indiana University and best known for his work on human sexual behaviour, was trained as a biologist, holding a chair in entomology and zoology.[31]

Canguilhem closes his lecture with a suggestion which has often been repeated.[32] It depends a little on knowing some Parisian geography.

> Thus it is with a degree of vulgarity that philosophy confronts psychology with the question 'Tell me what you are up to and I'll know what you are.' But once in a while, at least, the philosopher must be allowed to approach the psychologist as a counsellor and to say (once is not a habit): if you leave the Sorbonne by the exit in the Rue Saint-Jacques, you can either turn up the hill or go down towards the river; if you go up, you will get to the Pantheon which is the resting place of a few great men, but if you go downhill then you're bound to end up at the Préfecture de Police.
>
> (EHPS 381; WP 49)

Of course, giving this lecture at the Sorbonne, his audience would have fully known what he meant. The Rue Saint-Jacques runs roughly south-south-east from Île de la Cité and the Notre-Dame cathedral. It is a medieval street parallel to Haussmann's Boulevard Saint-Michel. The Pantheon and the police station are indeed along this road, and here Canguilhem's point is clear. But it is also important to realize that the first part of the description is significant too. The Sorbonne is a closed-off institution, with entrances from the street which open onto large courtyards, imposing buildings and an austere, scholarly status. It is not just a question of which direction psychology would take, but that this choice was first predicated on it leaving its academic solace.

One intriguing consequence was that it led to Hyppolite's suggestion that Foucault approach Canguilhem to be the *rapporteur* of his thesis on the history of madness in late 1958.[33] This lecture had another legacy too, as it was reprinted in *Cahiers pour l'analyse*, in 1966. Canguilhem's text was then read in a quite different context, shaped by Althusser's and Lacan's work, the reception of Foucault's ideas and the rise of antipsychiatry.[34]

The brain and thought

In 1980, again at the Sorbonne, Canguilhem returned to these questions in a lecture entitled 'The Brain and Thought'. This lecture continues the critique of psychology, and especially behaviourism, both in psychotherapy, and in some initial work in computing. Canguilhem suggests that, just as biologists think the study of the human brain cannot be done 'without situating it at the end of a history of living beings', he thinks a discussion of this topic needs to be situated 'within the history of culture' (BT 11/7). He quickly leads his audience from Aristotle's notion that the brain was for cooling the body to Hippocrates' suggestion that it was the base of sensations, movements and judgements, and Galen's experiments to prove this. He suggests the brain was seen as the seat of the soul, and spends a bit of time discussing Descartes, noting that traditional accounts completely misunderstand his argument about the pineal gland being the connection between soul and body. He locates the beginning of brain science proper with Gall and Spurzheim's work on the nervous system.[35] Even this was still tied down by the work of phrenology, which Canguilhem describes as 'that mixture of naivety and conceit ... essentially a cranioscopy grounded in the correspondence between content and container, between the configuration of [cerebral] hemispheres and the shape of the skull' (BT 12/7–8). One of Canguilhem's guides is Georges Lantéri-Laura's *Histoire de la phrénologie*.[36] He goes on to discuss some of the work to map the brain, with work on cerebral localizations. The details of his analysis are less significant than his telling remark that, since this led to work on lobotomies, 'we should pay attention to just how quickly apparent knowledge [*connaissance*] of the brain's functions was invested in techniques of intervention, as if the theoretical agenda was from the beginning driven by a practical interest' (BT 14/8).

In parallel to this work on neurology was the development of psychology as a science, though he suggested that it became 'little more than a pale reflection of physiology', encouraged by poor philosophy that simultaneously drew on that psychology. Taine is seen as one of the key figures here, and Canguilhem suggests that this work is not so distant. Canguilhem was in his mid-seventies when giving this lecture, and so already a figure of eminence to his doubtless much younger audience, but he reminds them it was only a few generations ago: 'the professors who taught my own professors, Bergson included', were determined to refute this work (BT 14/8–9). He similarly notes that the very young Freud was indebted to Taine's work, something which Canguilhem says Ludwig Binswanger noticed. Though Freud, of course, continued his intellectual development and criticized both psychic topography [*Topik*] and its relation to anatomy, these were still crucial elements of his work (BT 14–15/9).[37]

The relation of the brain to thought became an issue in the early twentieth century, and Canguilhem notes how two important French writers – Antonin Artaud and Raymond Roussel – both identified 'painfully lived experiences' with the brain. Pierre Janet, professor at the Collège de France and 'like Freud a student of Charcot' – challenged this assumption, saying that thought was solely rooted in the brain. It was a matter rather of behaviour as a whole, a corporeal process, and he claimed that 'psychology is the science of man as a whole, not the science of the brain' (BT 15–16/9).[38] Janet is also of interest to Canguilhem because he saw what were viewed as mental illnesses as actually social issues, a clash with social regulations: 'the word mad is therefore a designation [*appellation*] used by the police'.[39] Canguilhem thinks there are parallels here with contemporary antipsychiatry, and that Janet would have approved some Oxford student graffiti: 'Do not adjust your mind, there is a fault in reality.' Essentially, though, Canguilhem sees this as opening up the possibility of pursuing 'psychology without relying on arguments drawn from neurophysiology' (BT 16/10).

Canguilhem then discusses – through Descartes, the neurologist Sigmund Exner, Arthur Rimbaud, Nietzsche and Freud – the question of the subject. Are we justified in saying that there is a thinking subject, the 'I', that thinks, or rather is there a thought that occurs through a brain (BT 17–18/10)? This brings him to the question of what thinking is, which he recognizes has 'Heideggerian echoes', but wants to address from 'its banal and trivial angle' (BT 18–19/10–11). Here, he discusses Pascal and the idea of the brain as a computing machine, and the way modern computing built on some of the insights of nineteenth-century neurophysiology. The relation works both ways: 'we can speak of the computer as a brain or the brain as a computer' (BT 19/11).

Canguilhem gives a large number of examples, suggesting that, while computers may advance to more and more powerful capacity, there will always be something about thinking that eludes them. Invention and creativity seem to be beyond the machinic model. He quotes the mathematician René Thom: 'In this task, the human brain with its old biological past, its clever approximations, its subtle aesthetic sensibility, is still irreplaceable and will remain so for a long time.'[40] Computation models are found wanting, and so Canguilhem turns to work in cerebral chemistry. He notes that, despite antipsychiatry's resistance to psycho-pharmacology, there have been real advances in the treatment of conditions such as Parkinson's disease and schizophrenia. He wonders if this work might lead not to dealing with illness, but to improving brain performance. But, he suggests, there are limits here too – could such improvements work on issues such as invention, problem-solving or creativity?

Behaviourist work is also discussed by Canguilhem, though he notes that we should distinguish between variant forms of this more than is sometimes done: 'Pavlovian conditioning, which works by grafting a stimulus response relation onto an innate reflex relation; and Skinnerian or instrumental conditioning, which works, through repeated positive reinforcement, to consolidate forms of behaviour that achieve satisfactory solutions but that were initially discovered by chance' (BT 24/14). Yet, in all this work, there is a blurring of the line between 'learning and training [*dressage*]', a slip 'from the concept of education towards the concept of manipulation'. Even here, though, he recognizes that Pavlov's work was integrated into a dialectical materialist account which saw the 'human cultural environment' as 'a historical effect and not a natural given', and, as a consequence, 'thought is not a purely cerebral function, a biological product; it is a social effect relative to the type of society in which it intervenes'. In order to make sense of this, the context in which objects and phenomena in the environment appear to us is significant. This leads Canguilhem to the question of language: 'the thought–language relation refers to the brain–thought question via the language–brain relation' (BT 25/14). The issue of whether language is learned or innate is one which he traces through the then-recent debate between Jean Piaget and Noam Chomsky.[41] Canguilhem's view is clear – that human language needs to be understood as 'a semantic function' and that 'physicalist kinds of analysis' have been unable to explain it: 'To speak is to signify, to give to understand, because to think is to live within meaning [*sens*]. Meaning or sense [*sens*] is not a relation between …, it is a *relation to* … That is why it escapes every attempt to reduce it to an organic or mechanical configuration' (BT 27/15).

There is much more here, from discussions of Copernicus and Galileo to Ludwig Wittgenstein and Merleau-Ponty, and Paul Cézanne and René Magritte (BT 28–9/15–16). In the final part of the lecture, Canguilhem turns back to the process of thinking, which he suggests needs to be understood as 'a human practice that requires self-consciousness in presence to the world', though this should be understood not in terms of a privileged subject. Indeed, Canguilhem challenges the idea of interiority, which he suggests is 'exteriority turned inside out, but not abolished' (BT 29/16).

Canguilhem then returns to seventeenth-century debates between Descartes and Spinoza, siding more with the latter, and saying contemporary work must deal with the question of *ressort* – mind or spirit – which he discusses. Yet, in this, 'philosophy can expect nothing from the services of psychology', and he cites Husserl's suggestion that psychology has been a 'permanent disaster' since the time of Aristotle.[42] The problem with psychology is its attempt to be an objective science, but at the same time to 'situate itself within the other objective sciences and to teach them about the intellectual functions that allow them to be the sciences that they are' (BT 31/17). Philosophy can allow psychology to continue to make these kinds of suggestions, but its own task 'is not to increase thinking's output or yield [*rendement*], but to remind it of the meaning of its power' (BT 31/18).

Once more, Canguilhem's approach to these topics – of regulation and psychology – is resolutely historical. But it is historical in a philosophical register, because he is not concerned with the history simply for its own sake, but because of what light it can shed on present-day conceptual issues. Such concerns will continue through the quite different subject matter of his examinations of monstrosity, and especially the emergence of the understanding of evolution within the natural sciences.

6

Evolution and Monstrosity

From development to evolution

While we have a fairly good record of Canguilhem's lectures through his writings, even if relatively few of his courses were published, we have a much less complete record of his collaborative work and seminars. One exception is the seminar on evolution he led at the IHS in the 1958–9 and 1959–60 academic years. The results of this collaborative research were published as *Du développement à l'évolution au XIXe siècle* in 1962 as a special issue of the journal *Thalès*. Co-authored with Georges Lapassade, Jacques Piquemal and Jacques Ulmann, the study was reissued as a short book in 1985 and has gone through multiple editions since. The writing was initially done by the researchers separately, but each part was critiqued by and discussed with the others, and (with the exception of Canguilhem's brief foreword) the collective signed the whole text together. Canguilhem notes that a number of other people participated in the research seminar (DE 9). Ulmann went on to be well known for his works on medicine, education and sport,[1] and took over director-ship of the IHS on Canguilhem's retirement. Lapassade is best known for his sociological work on institutions and psychology, and his works of fiction.[2] Piquemal was one of Canguilhem's *lycée* students, and published little else: a study of the impact of the 1832 cholera epidemic on medical thought, an essay on Mendel, and some other short pieces, reprinted along with a lecture course on the history of respiration in a posthumous book on the history of medicine and biology, with a preface by Canguilhem.[3] Lapassade and Ulmann both died in 2008, Piquemal in 1990.

Du développement à l'évolution studies the century between two crucial works – Caspar Friedrich Wolff's *Theoria generationis* in 1759, and Darwin's *On the Origin of Species* in 1859 (DE 13).[4] Canguilhem notes that the centenary of Darwin was the initial spur, but that they were also intrigued by the concept of development, across a number of fields – from psychology to pedagogy, to politics and international planning, with the idea of under-development (DE 9). While Darwin's centenary was widely celebrated, the double centenary of Wolff was largely neglected, except in the USSR (DE 111).[5] On the opening page of the study itself, Canguilhem and his colleagues note that, between eighteenth-century naturalists and late nineteenth-century embryologists and zoologists, the terms 'development' and 'evolution' came to mean almost the opposite of their earlier use (DE 13). They add that, while there are a diverse range of uses of the terms, the biological is worthy of study to grasp the use of the concepts, independent of their wider usage (DE 13).

While the introductory essay discusses figures such as Aristotle, William Harvey, Malebranche and Leibniz (DE 14–15), the study as a whole is on the latter half of the eighteenth century through to the end of the nineteenth. While Canguilhem elsewhere is critical of the idea that the history of sciences can be reduced to biographies of scientists, that does in fact describe much of the structure of this study. The successive chapters treat Wolff; the relation between embryology and comparative anatomy; K. E. von Baer; Comte; Herbert Spencer; Darwin; Haeckel; John Fiske, Wilhelm Preyer and James Mark Baldwin; and Thomas Huxley and Charles Robin. Lamarck is a signal omission, but Canguilhem and his colleagues suggest that, while he is one of the 'founders of transformation … the *concept* of evolution, as far as his understanding includes a reference to the concept of development, owes little to him' (DE 109).[6]

Wolff and Darwin are two crucial figures, but they caution against seeing Darwin either as a figure without precedent or as a radical break. Indeed, they suggest that *On the Origin of Species* does not so much end the period initiated by Wolff's *Theoria generationis* as 'crown' it. It is the culmination of work challenging preformationism (DE 17). Wolff is a key moment to begin, they suggest, since his work initiated a number of concepts which Darwinism and post-Darwinism would further elaborate (DE 18). The discussion of Darwin is important, and in particular touches on his relation to work in geography (DE 64, 71), and on the concept of milieu (DE 65–6): 'Between Wolff and Darwin, we have seen a reversal of meaning has gradually imposed itself: the epigenetic signification seems definitely acquired with the liaison in 1859 between embryology and the transformatist conception of a kinship between species' (DE 79).

In their analysis, development and evolution are linked to the idea of generation, and reproduction (DE 13–14). In particular, they are interested in thinking about the relation between embryology and comparative anatomy (DE 27, 43). For earlier thinkers, 'the totality of the final form is determined by the totality of the initial form' (DE 15), a form in miniature. But this belief in preformationism was eventually eclipsed by epigenesis. The latter holds that organisms develop from a seed or egg in a complicated process of differentiation (DE 16). There are traces of epigenesis as far back as Aristotle, but there are many details to the story. Canguilhem and his colleagues spend some time on William Harvey, whose work on eggs in the seventeenth century in *On the Generation of Animals* was a point of tension between preformationism and epigenesis. Harvey's principle of *ex ovo omnia* should not be understood in a narrow sense of an egg, but in the sense of a seed, germ or foetus (DE 15). Epigenesis is in contrast with Biblical accounts of formation, in part because of the literal understanding of the time period between creation and the present. Famously, people like Bishop Ussher thought that the world was created in 4004 BCE, based on counting back from the birth of Christ with dates from the Hebrew Bible and other traditions.[7] This did not give enough time for gradual developments to have happened (DE 16–17; see 114–15). As Canguilhem and his colleagues claim at one point, 'all evolution is epigenetic' (DE 57).

Nonetheless, they realize that 'evolution' is not a straightforward concept and that its use can be differentiated:

> In the nineteenth century, the concept of evolution had a double sense, depending whether it is borrowed from positive philosophy established in the first half of the century, or if it borrowed from Darwinian biology and Spencerian philosophy established at the beginning of the second half. The fact that, towards the end of the nineteenth century, the popular [*vulgarisé*] concept of progress compounded these two meanings must not conceal that this alignment was in no sense systematic.
>
> (DE 53)

Yet this notion is broader than the specific sense that the term has: 'Evolution is not the privilege of biology. It asserts itself in all domains' (DE 56).

At around the same time as the seminar that led to this collaborative book took place, Canguilhem produced some other work on
Darwin, much of which was reprinted in *Études d'histoire*. In one
essay on Darwin and Wallace, he notes that 1958 was not just the
centenary of their work on the 'mechanism of biological evolution', but also the bicentenary of the tenth edition of Linnaeus's
Systema natura, which 'fixed the usage of the binary nomenclature'
(EHPS 99). He sees Darwin's importance, in part, as being a field
observer, outside of the university system, and as a contrast to the
'naturalists of the cabinet' who saw what he was doing as the work
of an amateur (EHPS 101). He suggests that Crowson is right to
see Darwin 'more as one of the last representatives of the naturalists of the eighteenth century than a forerunner of the group of his
successors of the twentieth century, the biologists of the laboratory'
(EHPS 102).[8]

One of the themes of this essay is the relation between Alfred
Russel Wallace's ideas and those of Darwin. This leads Canguilhem
to some reflections on precursors, which will be discussed in chapter
7 (EHPS 100–1). While the term 'natural selection' was not in Wallace's work, Darwin recognized that Wallace's work was operating
with the same ideas (EHPS 107). Yet Wallace was only looking at
the effects of adaptation, the way that animals and species adapted
to their milieu and how this best suited them. He also did not write
on sexual selection, and became hostile to this aspect of Darwin's
work (EHPS 107). Canguilhem suggests, in conclusion, that 'if Darwin
found in Wallace's writings the essential elements of his own ideas,
despite the absence of the term "natural selection", it is that this
term only designated the totality of certain conceptual elements in
his work' (EHPS 107). 'Natural selection' is not a force which is
added to the struggle for existence, but an element within it. Both
Darwin and Wallace recognized the importance of Malthus (EHPS
108–9), with Canguilhem suggesting that 'the common model to
Darwin and Wallace is Malthusianism, as an economic theory which
was both cause and effect of changes in the structure of English
society, with the replacement of agrarian capitalism with industrial
capitalism, under the imperative of free competition' (EHPS 109).

In another essay from this period, Canguilhem notes that, while
On the Origin of Species does not discuss human origins, Darwin had
thought about this question and held back some of his ideas on
the theme. He hints at this at the end of the book, but does much
more on the topic in the 1871 book *The Descent of Man*.[9] Canguilhem
notes Huxley's work in this area, and that of others (EHPS 113).[10]

In this work, he suggests, the 'genealogical tree of man' is substituted for 'the linear series of animals' (EHPS 116). In his striking phrase, for Darwin, 'nature was neither a theatre nor an artist's workshop: nothing is prepared, nothing is learned' (EHPS 116). The relation between the animal and man is a key question in this essay. Canguilhem stresses that, following Darwin, 'animality is the recollection of the pre-specific state of humanity; its organic prehistory and not its metaphysical anti-nature' (EHPS 116–17). This assessment took a while to impact on psychological work (EHPS 117–18), even though 'humans and animals possessed the same sensory organs, had the same fundamental intuitions, experienced the same sensations' (EHPS 119). Equally, he offers a challenge to work on non-human biology: 'Anthropocentrism is easier to reject than anthropomorphism' (EHPS 121).

As Balibar and Lecourt note in their 1985 preface to *Du développement à l'évolution*'s re-edition, the notion of evolutionism is understood as both 'science *and* ideology', in the sense that Canguilhem would elaborate in the later collection *Ideology and Rationality*.[11] They suggest, at that point, that, in the twenty-five years since the research, it had 'lost none of its topicality and relevance'. Indeed, they suggest that, to the multiple resonances of development that Canguilhem had noted, more might be added, ranging from biology to social and political ideologies.[12] Aside from this study, Balibar and Lecourt note the crucial work done in Canguilhem's other seminars at the IHS.[13] Among other events, these included the discussions of Cuvier which were published in *Revue d'histoire des sciences* in 1970. As well as Canguilhem, Francis Courtès, Dagognet, Foucault and Camille Limoges contributed essays to that project.[14]

Canguilhem returned to the themes of this engagement with Darwin in some essays included in *Ideology and Rationality*. In one, Canguilhem takes the example of the nineteenth-century ideology of evolutionism, with Spencer as case study (IR 42–3/36–7). He suggests that, while scientific work on evolution has changed since Darwin's time, his theories remain an 'integral part of the history of the constitution of the science of evolution'. However, he suggests that the 'evolutionist ideology is merely an inoperative residue in the history of the human sciences' (IR 43/37). As he explains in another essay included in this collection, the uses to which Darwin were put are responsible for this detour. The idea of the 'struggle for survival' was used in different ways, but it became 'political and ideological' in Spencer's usage. Any examination of the history of the life sciences needs to recognize that 'the theory of natural selection was initially regarded as an ideology by many people, not all of whom were fools' (IR 103/105). In that essay, 'On the History of Life Sciences since Darwin' (1971), Canguilhem begins by noting the parallels drawn by Freud:

> Sigmund Freud compared the scandal caused by the first public rev-
> elations of psychoanalytic theory to the early scandals provoked by
> Galilean cosmology in the seventeenth century and by Darwinian
> biology in the nineteenth century. And it is quite true that in all three
> cases man was stripped of a comforting illusion: that he occupied
> the centre of the universe, that his genealogical ancestry was unique,
> and the illusion of total access to his self-consciousness.
>
> (IR 101/103)

But Canguilhem says that, as a historian of science, there is some-thing worth examining in the 'sequence and relation of the first two of these disillusionments' (IR 101/103). He notes that not only did Newton's confirmation of Galileo's work fail to scandalize some theologians in England, but also they used it to argue against atheism: 'Thus the ultimate effects of the heliocentric cosmology, the theoreti-cal consequences of the first defeat of anthropocentrism, paradoxi-cally delayed and impeded the second, as a result of which man would be obliged to take his place as a subject in a kingdom, the animal kingdom, of which he had previously posed as king by divine right' (IR 102/104).

Canguilhem notes that the work of Charles Coulston Gillespie and Limoges has shown that 'the idea of a transformation of species by random adaptation to the constraints of the milieu based on individual reproductive differences was inconceivable until a previ-ous idea, that of a preordained adaptation of each species to its way of life, had been destroyed' (IR 102/104).[15] As he points out, 'the history of truth is neither linear nor monotone', and a revolution in one field does not necessarily lead to one in another. Cosmology does not lead to biology, and historical work can reveal how 'sci-entific discoveries in one field, if degraded into ideologies, can impede theoretical work in other fields' (IR 102/104–5).

The more general issue raised here is how subsequent work or practices may confirm earlier ideas. As Canguilhem suggests, Newton helped to confirm some of Galileo's theses, just as population genet-ics helped to confirm some of Darwin's ideas, in ways inconceivable at the time (IR 103/105). But he cautions against generalizing from this too much, and that the 'use of hindsight' risks 'making history seem straightforward and linear when in reality it was far more complex' (IR 103/105). Indeed, in general terms, the essay is concerned with showing how 'delay is as much a part of history, even the history of sciences, as progress' (IR 104/106). Most of the essay is devoted to analysing that delay – in this instance, between Darwin and the late nineteenth and early twentieth centuries.

Time was a crucial factor in embryology, and Darwin incorporated it into his analysis. Lamarck had done something with this idea, yet Darwin took it further.

> But the *Origin of Species* proposed a radically new idea, conceiving of time of life not as a power but as a factor whose effects could be perceived directly in distinct but complementary forms: the fossil was petrified time; the embryo, operative time; the rudimentary organ, retarded time. Together these bits of evidence constituted the archives of biological history, in which the biologist, by reading and making comparisons, could seek to establish a beginning. In the paleontological archive the beginning was the oldest fossil; in the embryological archive, it was the common element; in the morphological archive, it was the most rudimentary manifestation. Conceived in this way, the old comparative anatomy could be rejuvenated. The genealogical tree was the basis of the system, not the consequence. The common ancestor replaced the archetype. Classification ceased to be a static portrait of coexisting forms and became a vast synoptic canvas woven of the threads of time.
>
> (IR 106/108–9)

The gap was caused, Canguilhem suggests, by an absence of suitable understanding of the question of heredity. 'By what mechanism were variations inscribed in existing organisms, and how were they transmitted?' (IR 107/109). The techniques might have been there, but the hypotheses had not been formulated. On this question, Canguilhem holds to the line he took from Crowson in an earlier essay, suggesting that, while anticipating the twentieth century, Darwin 'remained a man of the eighteenth century' (IR 107/109):

> The historian of biology can see here a lesson of general import for the history of sciences. It was Darwin who said that 'in order to be a good observer one needs to be a good theorist'. The same thing can perhaps be said of the relation between practice and theory. Empirical data are theoretically useless without a theoretical specification of the conditions under which the data are valid and how they ought to be used. In order for science to make use of data collected in the course of a pre-existing practice, that practice must be translated into conceptual terms; theory must guide practice, not the

other way around. If it is true that the gap between theory and practice is often wide, it is no less true that no theory ever emerges from
empirical practice alone, in biology or in any other science.

(IR 107–8/110)

One interesting example comes in Canguilhem's discussion of
Mendel. If Darwin had died on the *Beagle* voyage, evolutionism
would have happened through the theories of others. Yet, while
Mendel proposed his theories not long after Darwin's famous works,
he was not known about for thirty years. What would have happened, Canguilhem wonders, if his papers from 1865 had not been
'preserved and catalogued without being read, or read without
being understood', but instead 'been destroyed by fire'? The laws
of genetics would be the same, but Mendel would be lost from the
story. And here there are two stories – the account of what happened
in public, where Mendel is not a known figure until this later date,
where 'discoveries' were actually rediscoveries; and the story of
the private development of the ideas, which genetics only later
caught up with. The historian of biology must therefore 'tell two
stories that only partially overlap' (IR 108–9/111–12).

Equally, what would have happened if Mendel's work had been
known earlier? Mendel cannot be understood as a precursor, who
anticipated work that others completed, because he 'followed the
trail all the way to the end'. Yet nor was he a founder, 'for surely
a founder cannot be someone unknown to those who erect an edifice
upon the foundations he laid'. Canguilhem suggests that, in the
absence of 'a pertinent category', a simile is perhaps the best we
can do: 'Mendel's scientific work was like a premature infant, which
died because the world was not ready to receive it ... a historical
pleonasm, at once embarrassing and beside the point' (IR 109–10/112).
This analysis of the 'counterfactual' is important. It is not to challenge the way we see the past, as much as to recognize the 'true
historical nature' of it, and to recognize the importance of individuals and their works, while seeing that they are situated within much
wider contexts and currents of thought (IR 110/112–13):

> the history of sciences of life is truly a history, that is, a series of rup-
> tures and innovations. I would use the word *mutations* if I were not
> averse to patterning the history of the life sciences after the history
> of life itself, and I would use the phrase *dialectical leaps* [*sauts*] if I
> were not afraid of being suspected of intellectual opportunism. In any
> case, the metaphor or model is of no importance. The important thing
> is to show not *who* did what but *how and why* the work was done.
>
> (IR 113–14/116–17)

These thoughts need to be situated within a wider discussion of Canguilhem's approach to history, which will be the focus of the next chapter.

Monstrosity and the monstrous

As this book's Introduction mentioned, Canguilhem had given a course on physiological abnormalities or monstrosity at the IHS in 1961–2.[16] He used some of the material in a lecture titled 'Monstrosity and the Monstrous' in Brussels on 9 February 1962, which was published in *Diogène*, and added to the second edition of *Knowledge of Life* the same year.[17] Here, Canguilhem returns to themes he had been exploring for many years. He begins with the claim that 'the existence of monsters calls into question the capacity of life to teach us order' (KL 171/134). Immediately, we might want to question what is meant by a monster, and it is worth noting that the French term *monstre*, and the related terms *monstreux* and *monstrosité*, can have a somewhat less charged sense relating to deformity. In addition, as is well known, the Latin root of these words, *monstrum*, comes from the verb *monere*, meaning 'to show' or 'to warn'. Indeed, this is key to the discussion that follows. The model of life is that the same engenders the same – a frog creates a tadpole which becomes a frog, flowers bloom on the plant, young animals are born from the same species. Where this model is broken, where there is a 'morphological divergence', where an other comes from the same, then we are 'gripped by radical fear' – not just a normal fear, but of a failure of life to work by its rules, and a risk that this could harm us, with a recognition it came from us (KL 171/134).

Monsters are always organic beings. There are no mineral or mechanical monsters; and later Canguilhem further qualifies this to say they must be living. While we might use the term 'monstrous' to describe something else, such as a mountain, this is a mythical or emotive sense. As such, Canguilhem distinguished the monstrous from the enormous. The latter, literally the e-normous – that is, a break from a norm, an excess – is excessive only by a metric, an order of magnitude. Nonetheless, a giant would render even this distinction problematic – 'beyond a certain degree of growth, quantity calls quality into question' (KL 172/135). 'The monster is a living being with negative value', and Canguilhem stresses that it is monstrosity, not death, that is the real counter-value to life (KL 172/135). Yet, as well as fear, monstrosity provokes 'curiosity, even fascination': it is seen as marvellous, even if inverted (KL 173/136).

Monstrosity provokes many thoughts. Life, and its reproduction, is not as straightforward as we imagined; but equally, since monstrosity is a failure, all of life's successes are the avoidance of that failure. Following Bachelard, Canguilhem suggests that 'the true is the limit of lost illusions' (KL 173/136),[18] and quotes Gabriel Tarde who said that 'the normal type is the degree zero of monstrosity' (KL 173/136; see 160/126).[19] Perhaps it is even worth underlining the relative rarity: 'Life is poor in monsters. The fantastic is a world' (KL 173/136; see 183–4/145–6).

The 'relation between the monstrosity and the monstrous' is, Canguilhem suggests, 'thorny'. Etymologically related, 'they are in the service of two forms of normative judgment – the medical and the juridical – which were initially confounded rather than combined in religious thought, and then progressively abstracted and secularized' (KL 174/137). In antiquity and the Middle Ages, monstrosity was seen as an effect of the monstrous. The monstrous was associated with the criminal, even the diabolical, and the hybrid was a specific focus – 'an infraction of the rule of the sexual segregation of species and a sign of a will to perversion in the table of creatures. Monstrosity is a consequence less of the contingency of life than of the licentiousness of the living' (KL 174/137). Historically, with humans, different cultures have worked in different ways – 'if the Orient deifies monsters, Greece and Rome sacrifices them' (KL 174/137). Canguilhem gives various examples, of women and devils, of birthmarks, of animals which resemble different species, and draws on Ernest Martin's *Histoire des monstres* in particular.[20] At a certain moment, 'the monstrous, initially a juridical concept, was gradually made into a category of the imagination' – the link between teratology and demonology is crucial to Christian theology (KL 175/138). Teratology is, today, the study of abnormalities, but its earlier sense was the mythology of the fantastic and the monstrous – a secondary meaning today.

The shift from teratology being concerned with monstrosity, in its specific forms, rather than the monstrous in its general sense, comes about in the nineteenth century. While formerly it could encompass myths and legends – and Canguilhem goes into some detail here – now it wishes to categorize, order and organize. While much of what Canguilhem says here trades on *The Normal and the Pathological*, which inspired the early work of Foucault, he now makes use of Foucault's *History of Madness* to stress his point:

> The same historical epoch that, according to Foucault, naturalized madness occupied itself with naturalizing monsters. The Middle Ages

(which is not so named because it allowed the co-existence of extremes) is the age in which one sees the mad living in society with the sane and monsters with the normal. In the nineteenth century, the madman is in the asylum, where he serves to teach reason, and the monster is in the embryologist's glass jar, where it serves to teach the norm.

(KL 178/140)[21]

Yet this is not a stark transition from one mode to another. The eighteenth century, Canguilhem says, 'was not too hard on monsters'. While some were undoubtedly indicted or expelled, others were used as a means to make sense of regular phenomena. Leibniz called them 'intermediate creatures' – their equivocal nature helped to reveal the transition between species (KL 178/141):[22] 'In sum, whether in embryology, classification, or physiology, the eighteenth century made the monster not only an object but an instrument of science.' It is the nineteenth century that truly sees the shift – 'the scientific explanation of monstrosity and the correlative reduction of the monstrous' (KL 179/142).[23]

This then is the transition to a modern understanding: 'From this point on, constituted by descriptions, definitions, and classifications, teratology became a natural science.' Biology itself was a relatively new term and concept, only preceding teratology briefly, and both became experimental sciences. In the middle of the nineteenth century, Camille Dareste 'founded teratogeny, the experimental study of the conditions for the artificial production of monstrosities … Echoing Marcelin Berthelot, who said that chemistry creates its object, Dareste proclaimed that teratogeny must create its own objects' (KL 180/142–3).[24] Reprising and developing themes from *The Normal and the Pathological*, Canguilhem proclaims:

> From then on, monstrosity appears to have revealed the secret of its causes and laws, while anomaly appears called upon to explicate the formation of the normal. This is not because the normal is an attenuated form of the pathological, but because the pathological is the normal impeded or deviated. Remove the impediment and you obtain the norm. Henceforth, the transparency of monstrosity to scientific thought cuts monstrosity off from any relation to the monstrous.
>
> (KL 180–1/143)

As his essay draws to a close, Canguilhem sketches how realism in art begins to erase the monstrous, how 'positivist anthropology set out to depreciate religious myths and their artistic representations', and how perhaps 'the monstrous seeks refuge in poetry' (KL

181/143). But he wonders whether it is in the creation of monstrosity in science, through biological experiments, that we should seek the monstrous today: 'the ignorance of the ancients held monsters to be games of nature, while the science of our contemporaries turns them into the games of scientists ... It would thus be the century of positivism that would have realized – thinking it was abolishing it – what the Middle Ages had dreamt' (KL 182/144).

Returning to his earlier claim, Canguilhem concludes that 'we can now understand why life is relatively poor in monsters: it is because organisms are incapable of structural eccentricities except during a short moment at the beginning of their development' (KL 183/145). The fantastic is a world, in the second half of his phrase, because it is conditioned by the imagination, whose power is 'inexhaustible, indefatigable', due to the fact that 'the imagination is a function without an organ ... it feeds only on its own activity' (KL 183/145). This is the first of two reasons for keeping the duality of monstrosity and the monstrous: the 'poverty' of monsters, of monstrosity; but the 'prodigality' of the monstrous. The second reason is that there is 'nothing monstrous about monstrosities', because monstrosities, in an age of positivism, are not exceptions to nature, not outside its laws. Its deviations are deviations within, rather than beyond. But the monstrous holds to a different logic, 'a chaos of exceptions without laws', an 'imaginary, murky and vertiginous world' (KL 184/146).

The day after this lecture, Canguilhem gave another talk in Brussels, titled 'On the Singular and Singularity in Biological Epistemology' (EHPS 211–25). It also related to themes from his teaching – in this case, from a few years before.[25] There is little overlap between the two Brussels lectures, save for some brief discussion of monstrosity in the second (EHPS 219–20). A detailed reading of this lecture will be forgone here, but it is an intriguing piece which sketches a very cursory history of zoology from Aristotle. One key claim is the dual sense of singular as both unique and isolated: 'Buffon recognized the two functions of this adjective, exclusive and separate [partitif], qualitative and quantitative. The singular is such because it is different from all others; the singular is such because it is separate' (EHPS 214). This matters, because, at the end of the seventeenth century, Canguilhem suggests, classificatory botany and zoology had a sharp contrast, with the morphological singularities of the animal kingdom compared to the 'vegetal world, taken as a whole, which was presented as a singularity' (EHPS 216). Some of the work he was doing in his seminar, to be discussed in the next chapter, concerned how that situation changed.

* * *

On 23 and 24 February 1966, Canguilhem returned to Brussels to give two lectures under the general title of 'Le concept et la vie' [Concept and life]. They were published in *Études d'histoire* in 1968. In the lectures, Canguilhem suggests that 'the nature and the value of the concept are in question here, as much as the nature and the sense of life' (EHPS 335). Yet Canguilhem is intrigued why there might be a complicated relation between concept and life, 'for the theory of the concept and the theory of life have the same age and the same author'. That author, of course, is Aristotle, 'the logician of the concept and the systematic philosophers of living things' (EHPS 336; VR 303). Indeed, he takes one of Aristotle's celebrated descriptions of the human as if it were a naturalist's – not the *zōon politikon*, or political animal, but the *zōon ekhon logon*:[26] 'If the definition of man as *zōon logikon*, or reasoning animal, is a naturalist's definition (in the same sense that Carolus Linnaeus defines the wolf as *canis lupus* or the maritime pine as *pinus maritima*), then science, and in particular the science of life, is an activity of life itself' (EHPS 337; VR 304).

These lectures are wide-ranging and discuss thinkers including Buffon, Linnaeus, Kant, Hegel, Bernard, Goldstein and Bergson, through to James Watson and Francis Crick's work on DNA. Canguilhem ends these lectures with a sequence of questions that cycle back to the themes of the book *Knowledge of Life* and recur through his work as a whole:

> What, then, is knowledge [*connaissance*], because we must certainly finish with this question? If life is concept, does recognizing the fact give the intelligence access to life? What, then, is knowledge? If life is meaning [*sens*] and concept, how do we conceive of the activity of knowing ... Most scientific techniques, it can be argued, are in fact nothing more than methods for moving things around and changing the relations between objects. Knowledge, then, is an anxious quest for the greatest possible quantity and variety of information. If the *a priori* is in things, if the concept is in life, then to be a subject of knowledge is simply to be dissatisfied with the meaning one finds ready at hand. Subjectivity is therefore nothing other than dissatisfaction. Perhaps that is what life is. Interpreted in a certain way, contemporary biology is thus a philosophy of life.
>
> (EHPS 364; VR 318–19)

Chapters 4, 5 and 6 of this study, therefore, have treated his multiple studies within this vast topic of the philosophy of life. They build on his interest in the normal and the pathological, broadened through his 1947 lectures on the philosophy of biology, discussed in chapters 2 and 3, respectively. Throughout this period, crucially in the 1950s and 1960s, but extending to his occasional writings of the 1970s and 1980s, Canguilhem was continually interested in the relation of philosophy to the life sciences, worked through his general interest in physiology, regulation and evolution and more specific studies of the reflex, psychology and monstrosity. The readings here have remained close to his texts, organizing and relating their concerns rather than trying to systematise. In all these studies, while he raises many wider scientific and philosophical issues, Canguilhem always explores issues historically. In his discussions of the history and philosophy of the life sciences, Canguilhem both draws on earlier figures who studied quite different sciences, and outlines some general principles which structure how such work might be done by others. His contribution to the philosophy of history is the topic of the next chapter.

7

Philosophy of History

Jean Cavaillès

A figure much remembered by Canguilhem is the philosopher, logician and mathematician Jean Cavaillès.[1] Cavaillès was the year ahead of Canguilhem at the ENS, and his predecessor in the post at Strasbourg. He was executed by the Gestapo in 1944, and Canguilhem later wrote a number of short tributes. Canguilhem thought very highly of him, though there is a strong sense that his subsequent reminiscences are shaped by Cavaillès's fate.[2] The short book *Vie et mort de Jean Cavaillès* includes texts from 9 May 1967 for the inauguration of the Jean Cavaillès lecture theatre at Strasbourg; a radio address for France-Culture on 28 October 1969; and a thirtieth anniversary event on 19 January 1974 at the Sorbonne, where the Salle Cavaillès was being opened.[3] Some earlier texts are included in *Oeuvres complètes*, vol. IV.

After his *agrégation*, Cavaillès taught in the *lycée* at Amiens, and went on to obtain a doctoral degree in 1938. As well as his formal French education, Cavaillès spent time in the 1930–1 academic year in Germany as a Rockefeller exchange student. There he attended lectures including ones by Heidegger in Freiburg, witnessed the political turmoil, and in March 1931 heard Hitler speak in a Munich hostelry.[4] He read *Mein Kampf* in the early 1930s (VM 11, 15, 34). He was influenced by meeting Husserl and by hearing the lectures in France which were the basis of *Cartesian Meditations* (VM 27).[5] He also worked on the manuscripts of Georg Cantor, the founder of set theory (OC IV, 255). Canguilhem notes that 'he always read,

studied and we can almost say practised Spinoza', finding him
much more appropriate for his work than Malebranche or Leibniz
(VM 26, see 42; OC IV, 270). While a mathematician, he was also
'sustained by poetry', citing Rimbaud, among others, in his work
(VM 30). He was appointed to a position at Strasbourg in 1938.

With the outbreak of war between France and Germany, Cavaillès
joined up with an infantry unit and took part in the initial fighting,
and was captured in 1940. After the French surrender, he escaped
and joined his Strasbourg colleagues in exile in Clermont-Ferrand.
Canguilhem recalls that Cavaillès was appalled by the surrender
and continued to fight, suggesting that while many French were
part of the resistance, Cavaillès was one of those who made it (OC
IV, 220, 257). His work for the resistance took many forms, including liaison visits to London, military operations, transporting explo-
sives and so on (VM 46). Along with Emmanuel d'Astier de la
Vigerie, he was one of the founders of La Dernière Colonne (The
last column) – which later became Libération-Sud (Liberation-South)
– and its newspaper *Libération* (OC IV, 221, 257–8; VM 11, 19, 45).
His call to the Sorbonne in 1941 did not end his work for the resistance, as he became a founder of Libération-Nord (Liberation-North),
and Canguilhem wonders whether the 'Salles Cavaillès' at the Sorbonne should not bear the name of his multiple pseudonyms from
this time (VM 45–6).

Aside from a few articles, Cavaillès's works are included in *Œuvres
complètes de philosophie des sciences*.[6] His writings were technical works
of mathematical theory for the most part, discussing set theory,
axioms and formal method.[7] Cavaillès's *Oeuvres complètes* also
includes Canguilhem's three tributes from *Vie et mort*, along with
the transcription of a radio broadcast from 1989.[8] This volume also
includes the posthumous *Sur la logique et la théorie de la science*, co-
edited by Canguilhem in 1947, for which Bachelard added a preface
to the 1960 second edition.[9] This work had been written in 1942
while Cavaillès was in captivity in Montpellier (VM 28). Canguilhem
notes that, for a philosopher, preparing for death, it is customary
to write an ethics. Cavaillès chose to write a logic: 'He thus gave
us his ethics, without having written it' (VM 29). Elsewhere, Canguilhem suggests that it was his logic which unified Cavaillès the
philosopher and the resistance fighter (OC IV, 225). Tellingly, the
final lines of *Sur la logique et la théorie de la science* declare that 'it is
not a philosophy of consciousness but a philosophy of the concept
which can provide a doctrine of science. The generative necessity
is not an activity, but a dialectic.'[10] Cavaillès did escape Montpellier,

and spent some time in London, meeting with de Gaulle (VM 21), as well as returning to teaching in Paris.[11] Canguilhem recalls that they met for the last time in May 1943 (VM 21; OC IV, 905–6). After leading attacks on the German naval base in northern France, Cavaillès was betrayed and arrested in August 1943. He was tortured and imprisoned before being executed in February 1944.[12] His body was thrown in a mass grave, and when discovered was originally given the designation 'unknown #5'. Canguilhem suggests that both terms would have special resonance for a mathematician (VM 21–2; OC IV, 906–7). He was later reburied at the Sorbonne.

In his multiple tributes, Canguilhem tends to focus on the story of the life and his resistance, without as much attention being given to the work. This is certainly not because of a lack of interest or recognition of its importance: his editorial work on posthumous publications gives one indication of that. In his longest tribute, from 1947, Canguilhem does go into some details about Cavaillès's work on set theory and how it was an intervention in contemporary debates on the topic (OC IV, 264–9). Some of the discussion is particularly technical. But the fundamental point that Canguilhem seems to take is his claim that 'Cavaillès addressed his problem with a historical approach' (OC IV, 264). Cavaillès was both a philosopher, and a historian, of science.

Gaston Bachelard

In this philosophical-historical approach, Cavaillès's work is operating in a similar register to Bachelard's work on physics and mathematics. It is not surprising that Canguilhem notes that Bachelard is the inspiration for thinking about this relation between epistemology and the history of sciences, but he suggests that many contemporary readers have come to his work second-hand, through commentaries on his work, especially by Lecourt (IR 20–1/10–11). For Canguilhem – though he was always supportive of Lecourt, and helped to get his early work published – he thinks that Bachelard is his own best introduction: 'to my mind, the best summary of Bachelard's research and teaching can be found in the concluding pages of his last epistemological work, *Le Matérialisme rationnel*' (IR 21/11).[13]

Bachelard's work is enormously wide-ranging, from works on the philosophy and history of science through to his later works on the elements, poetry and space. However, unlike Canguilhem himself, the bulk of his writings were in book form. As Canguilhem notes in a preface, the posthumous collections *L'engagement ration-aliste, Études* and *Le droit du rêver* include 'almost all of the writings of Gaston Bachelard outside of his books'.[14] This is not the place for a sustained discussion of Bachelard, but, before turning to Canguilhem's approach, it is worth spending a little time on him.[15] This will mainly be done through the lens of Canguilhem's own writings on his mentor.

When Bachelard died in 1962, Canguilhem wrote a number of tributes to him. In one, he claims that the history of sciences did not have the same status in France as in other countries, especially in the teaching of advanced courses, but there was an association with work on the philosophy of science. Bachelard's appointment in 1940 to the Sorbonne, for a position of teaching the history and philosophy of sciences, was therefore a significant moment (EHPS 173). Before Bachelard, the dominant approach in France had been the positivist school, of August Comte and his followers. The Sorbonne post was initially in the history of philosophy and its relation to the sciences, which became the History and Philosophy of Science position that Bachelard, and, from 1955, Canguilhem, would occupy (EHPS 173–4).

Bachelard arrived in Paris for this position with an extensive range of published books – ranging from the studies of the scientific spirit to *The Psychoanalysis of Fire* and his works on time, *The Dialectic of Duration* and *The Intuition of the Instant*.[16] But Canguilhem singles out his two theses as fundamental to the work he would do, especially on the relation between the history and philosophy of the sciences. These theses were *Essai sur la connaissance approchée* and *Étude sur l'évolution d'un problème de physique*.[17] The first was an epistemological study, examining the notions of reality and truth; the second traced the formation of scientific concepts. Canguilhem suggests that this work was crucial for 'a conception of the history of sciences' and sat in relation to the philosophy of science, with the crucial 'concept of the *epistemological obstacle*'. This was a critique of a certain way of writing the history of science. It particularly challenged the idea of simple progression, with the idea of 'progressive complication, and the misunderstanding of the tenacity of errors which long obscured a problem' (EHPS 176). Bachelard's 1938 book *The Formation of the Scientific Mind* showed in its first chapter that the root of these errors 'should be sought within the scientific knowledge itself, rather than outside it' (EHPS 176; see 167).

Canguilhem similarly suggests that history had tended to neglect the history of sciences. Today, science and technology studies has somewhat challenged that neglect, and work in this area is much more common. But in the middle of the twentieth century, Canguilhem claimed that 'general history is above all political and social history, completed by the history of religious and philosophical ideas' (EHPS 10). He added that even a history of society which accounts for 'juridical institutions, economy and demography' often does not include 'the history of scientific methods and theories' (EHPS 10). Philosophers are generally led to the history of science only if they are interested in the history of philosophy, and scientists often know little about the background to their work.

Bachelard's work on the history of science was mainly concerned with physics, and that in its particularly mathematical form. Indeed, Canguilhem suggests that, for Bachelard, mathematical physics was the 'royal science' (EHPS 175), and his understanding of mathematics is in contrast to that of the logical positivists. Bachelard contended that mathematics did have epistemological content, and that progress could add to this. This was a position he shared with Cavaillès, whose critique of logical positivism is one that Canguilhem himself endorses (IR 23/13; see EHPS 175). Canguilhem recognizes, though, that Bachelard was not really interested in the history of biology (EHPS 170), and that his model best fits mathematical physics and nuclear chemistry. It might seem that Bachelard's approach would not work so well for natural history, but Foucault rightly acknowledged the importance of Canguilhem's shift of the history of sciences from the heights of mathematics, physics and astronomy to the sciences of life.[18] As has been stressed, it is important that the specific phenomenon of life is not lost in a reductive application of other scientific approaches. In particular, Canguilhem contends that 'the recursive method must be used judiciously, and we must learn more about ruptures' (IR 24–5/14–15). He argues that it is not really a flash of insight or genius to suggest a shift, but actually it is often 'a series of successive or partial ruptures. A theory is woven of many strands, some of which may be quite new while others are borrowed from older fabrics' (IR 25/15).

Bachelard was also significant for Canguilhem because he was 'the first French epistemologist who thought, wrote and published, in the twentieth century, at the chronological and conceptual height of the sciences he dealt with' (EHPS 185; see OC IV, 729–39). The most effective way to see both Bachelard's and Canguilhem's practice as historians is, of course, in action. Canguilhem's multiple histories of medicine and biology have already been discussed in detail. In those histories, Canguilhem shows a deep knowledge of the history of the physical sciences more generally. At times, this is his explicit focus, though in the history of mathematics, physics and chemistry he generally defers to Bachelard, Koyré and Cavaillès. This means that Canguilhem mainly concentrates on the history of the life sciences, and only occasionally discusses the physical sciences.

Canguilhem contributed to a 1964 tribute volume to Koyré, contributing an essay on 'History of Religions and History of Sciences in Auguste Comte's Theory of Fetishism' (EHPS 81–98).[19] He is generally appreciative of his work, seeing the parallels with Bachelard's, and suggesting that, in relation to the physical sciences, Koyré has treated these questions 'in a decisive way' (EHPS 41, see 43). Canguilhem has in mind *The Astronomical Revolution* and his *Galileo Studies*:[20] 'Galileo, we can say with Alexandre Koyré, was in the true [*dans le vrai*]' (EHPS 46). This is a phrase he elsewhere attributes to Bachelard, and uses frequently. As he clarifies, 'being in the true does not mean to say always true' (EHPS 46). Rather, as becomes clear in some of his other writings, it is more about being within the realm where a judgement of true or false is even possible.[21] He was also fond of quoting Koyré's comment that 'the history of science is certainly not a dead history. Nevertheless it is, *grosso modo*, the history of dead things.'[22]

Canguilhem says relatively little about Bachelard's more poetic work. But he does note that, while he was 'indulgent to poets and painters, Bachelard was demanding of philosophers … in his epistemological work, the 'philosopher' is typically, even if sometimes lightly, cartoonish [*caricaturel*]: he plays the role of a bad pupil in the school of contemporary science, a pupil sometimes lazy, sometimes absent-minded, always behind the idea of the master' (EHPS 187). While the particular mode of engagement of these pieces, which are effectively obituaries, should not be forgotten, Canguilhem is clear in his stress of Bachelard's originality and importance: 'If the history of the sciences consists of counting the variants in the successive editions of a Treatise, Bachelard is not a historian of the sciences. If the history of the sciences consists of making perceptible [*sensible*] – and also intelligible – the difficult, frustrated, repeated and corrected edification of knowledge [*savoir*], then Bachelard's epistemology is always a history of the sciences in action' (EHPS 178).

Bachelard put the history of sciences up as 'a philosophical discipline of the first rank … he did more than pave a way, he fixed a task' (EHPS 186). In order to do justice to his work and memory, Canguilhem suggests, we should not just acknowledge the loss, but make sure that the 'lesson of this man of genius' is not lost (EHPS

186). Canguilhem's most explicit reflections came later in his career, especially in the 1960s and 1970s. Nonetheless, these discussions of Bachelard's and Koyré's importance help to situate the work he developed from the 1940s onwards. While both Bachelard and Canguilhem combined history and philosophy, Bachelard was primarily a philosopher of science; Canguilhem was much more historical in his approach.[23] Bachelard's 'epistemology was historical'; Canguilhem's 'history of the sciences is epistemological'.[24] The distinctions between their work, and its development by Foucault and others, are perhaps masked by the general phrase 'historical epistemology'. Canguilhem's debt does not mean that his work is a simple transformation of Bachelard's approach into different areas.[25]

The history of sciences

Canguilhem regularly talks about the history of sciences in the plural. This sounds odd to the English ear, and it might imply that he is proposing a simple relativism. But his point is more that we should not put all different sciences into a single term of 'science'. His usage is closer to the way in English we talk of the physical sciences, life sciences or human sciences – a range of different disciplines and approaches that often share some methodological presuppositions and generally some object of analysis. Canguilhem frequently reflected on these issues. One example is 'Cell Theory in Biology: The Meaning [*Sens*] and Value of Scientific Theories', delivered as a lecture in 1945, published in 1946 and reprinted in *Knowledge of Life* simply under the title of 'Cell Theory'. The most complicated word in the original title is *sens*, which is sometimes simply 'sense', but also 'meaning' and, crucially, 'direction'.[26] Some of the biological aspects of this lecture are discussed elsewhere, and its broad history of cell theory is too specialized and too detailed to be summarized here. Nonetheless, there are some important claims about the history of sciences made in this piece which are worth discussing here, beginning with the claim that it is a neglected field:

> Its very meaning [*sens*] is in flux. Should the history of sciences be written as a special chapter in the general history of civilization? Or should one look in its scientific conceptions for the expression of an age's general spirit, a *Weltanschauung*? Problems of attribution and competence also remain unresolved. Is this history the province of the historian, with his competences as exegete, philologist, and erudite

(especially for antiquity) or rather of the specialized scientist, capable
of mastering, in his capacity as expert, the scientific problem whose
history he retraces?

(KL 43/25)

The division of academic labour here is important. The history
of sciences is a specialist field, and requires some degree of scientific
knowledge, rather than seeing it as merely an expression of a particular space and time. Science, for Canguilhem, is not an epiphe-
nomenon, but nor is it the key to understanding beyond its own
scope. But his last question raises a significant issue. What is the
relation of the history of sciences to the sciences themselves? 'If the
true [le vrai] – the goal of scientific research – is exempt from historical transformation, then is the history of science anything more
than a museum of errors of human reason?' (KL 43/26). Such a
conclusion would have a major impact: 'In that case, the history of
sciences would not be worth a single hour's effort for the scientist
[le savant], for the history of sciences would be a question of history
but not of science … more of a philosophical curiosity than a stimu-
lant to the scientific spirit' (KL 43/26).

Understandably, Canguilhem suggests things are more complicated, arguing that this kind of attitude presupposes a dogmatic
view of science and scientific critique, a belief in the 'progress of
the human spirit', 'the mirage of a "definitive state" of knowledge
[savoir] … Something is considered an error because it is from yesterday' (KL 44/26). He describes such a view in a pithy phrase:
'Chronological anteriority is taken for logical inferiority' (KL 44/26).
This is a position Canguilhem attributes to Bernard, in particular
(KL 44 n. 2 / 160 n. 2), and seems to have in mind this claim: 'No
experimental science, then, can make progress except by advancing
and pursuing its work in the future. It would be absurd to believe
that we should go in search of it in the study of books bequeathed
to us by the past. We can find there only the history of the human
mind, which is quite another matter.'[27] The history that we therefore
find in positivist accounts is often a history of prescientific thought,
of myths rather than science: 'According to that conception of science,
and despite its equation of the positive with the relative, the positivist notion of the history of sciences masks a latent dogmatism
and absolutism' (KL 44/26).

Canguilhem gives a couple of examples. In the science of optics,
the theory of waves and the theory of emissions were synthesized
in wave mechanics, and this, rather than one eliminating the other,

helps to make sense of developments. Newton recognized the limitations of his theory of emission, and it was his disciples that turned his thoughts into a dogmatic view which made its modification difficult (KL 44–5/26–7). In biology, the problem of species illustrates a similar point. While Linnaeus is often seen, and criticized, 'as the authoritarian father of the theory of fixed species', his own writings indicate that he understood possibilities of transformation, through his experiments on hybridization: 'Linnaeus's meditation on monstrous and "abnormal" varieties in the animal and vegetable kingdom led him to abandon completely his first conception of species' (KL 46/28). So Linnaeus, through the successive versions of his *System of Nature*, changed formulations, and 'never came to a clear notion of species' (KL 46/28). One reading of his work can justify a theory of fixity; but, on the basis of his work as a whole, a different reading could be justified. 'It seems to us that the benefit of a history of sciences properly understood is to reveal the history in science – by which we mean the sense of possibility' (KL 47/28):

> It is thus profitable to look for the elements of a conception of science and even of a method of culture in the history of sciences, understood as the psychology of the progression by which notions have attained their current content, as the articulation of logical genealogies and – to use an expression of Gaston Bachelard – as a census of 'epistemological obstacles' overcome!
>
> (KL 47/29)[28]

One of the themes of this lecture is the crossover between some sciences and others. Canguilhem suggests at one point that 'it is incontestable that Buffon sought to be the Newton of the organic world, a bit like Hume, in the same period, sought to be the Newton of the psychic world' (KL 55/35). The parallel is not spurious: Buffon had translated Newton's *Method of Fluxions*, to which he added a preface (KL 54–5/35). Technological advances and uses are also significant. To compare the relation between Bichat and Pinel, and the shift from tissue to cell, Canguilhem comments that this was partly about the instruments they used. 'Bichat did not like microscopes, perhaps because he did not know how to use them well … Bichat preferred the scalpel … But at the tip of a scalpel it is just as impossible to discover a cell as it is to find a soul. We intentionally allude to this materialist claim' (KL 64/43).[29]

Canguilhem is not trying to say 'there is no difference between science and mythology, between measurement and reverie', but equally and 'inversely, to want radically to devalorize old intuitions on the pretext of their theoretical obsolescence renders one – imperceptibly but inevitably – unable to grasp how such a stupid humanity could one fine day have woken up intelligent' (KL 80/56). Canguilhem recognizes that many scientists were well aware of these issues, and that some of them were able to recognize the 'obstacles and limits of this theory'. This included those who made the most progress in developing the modern view. He suggests that the need to develop a more nuanced understanding of intellectual development in the history of sciences will not be a surprise to scientists, but really only to those whose understanding of how knowledge actually develops is lacking. Without this kind of understanding, 'there can be neither scientific critique nor a future for science' (KL 80/56).

Epistemology and the history of concepts

Canguilhem makes it clear that we should not confuse the history of physiology with the history of physiologists (EHPS 232; VR 105). Although much of his work does appear to be the sequence of key figures, and their influence and critique, he is much more concerned with concepts. One of the implications of the new scientific spirit was a new way of writing the history of sciences[30] 'This history could no longer be a collection of biographies, nor a table of doctrines, in the manner of a natural history. It must be a history of conceptual filiations. But this filiation has a status of discontinuity, just like Mendelian heredity. The history of sciences must also be demanding, as critical as is science itself' (EHPS 184). It 'should be a history of the formation, deformation and rectification of scientific concepts. Since science is a branch of culture, education is one of its conditions of invention. What the individual scientist is capable of depends on what information is available, if we forget that, it is easy to confuse experimentation with empiricism' (EHPS 235; VR 110–11). Concepts are crucial, and this is an important caution against what might be called a naïve empiricism. As Canguilhem stresses: 'if science were purely empirical it would be impossible to write its history, since it would be the history of random events' (EHPS 236; VR 111).

Canguilhem would further reflect on these questions in his introduction to Études d'histoire: 'There are three reasons for doing the history of sciences: historic, scientific and philosophical' (EHPS 11). Scientific reasons include knowing enough of the tradition to recognize the basis for new theories or experimental results; historical ones concern questions of priority, paternity and commemoration. As Canguilhem notes, 'the history of sciences abounds in quarrels about priority' (EHPS 257). We might think of the argument about whether Leibniz or Newton first discovered calculus – a dispute which has gone through multiple stages, from one involving the figures themselves or their surrogates, to later work on manuscripts and correspondence, and what seems to be the current consensus that they arrived at the work independently. Newton seems to have devised the infinitesimal calculus; Leibniz, the differential and integral calculus, and the annotation used today.[31] In terms of the philosophical importance, Canguilhem is following Eduard Jan Dijksterhuis's suggestion that 'the History of Sciences forms not only the memory of science but also its epistemological laboratory'.[32] Dijksterhuis continues, in a line not quoted by Canguilhem: 'It not only recalls the work of the predecessors without whose exertion and ingenuity our present-day science would not exist, but also makes it clear what course had to be followed in order to make it possible.'[33]

Canguilhem notes that Pierre Lafitte, one of Comte's disciples, compared the history of science to a 'mental microscope',[34] but Canguilhem has doubts about the positivist bias of this formulation, with the idea that the 'historian's object is lying there waiting for him' to discover it, and that 'all he has to do is look for it, just as a scientist might look for something with a microscope' (EHPS 12–13; VR 43). He also doubts that 'the history of sciences is merely to the sciences like an apparatus of detection is to already constituted objects' (EHPS 13). This is one of the many reasons that he thinks epistemology can be of use to straightforward history:

> Epistemology provides a principle on which judgement can be based: it teaches the historian the language spoken at some point in the evolution of a particular scientific discipline, say, chemistry, and the historian then takes that knowledge and searches backward in time until the later vocabulary ceases to be intelligible, or until it can no longer be translated into the less rigorous and more common lexicon of an earlier period.
>
> (EHPS 13; VR 43)

When Canguilhem takes this approach, epistemology allows him to make a crucial distinction between two ways we can see the history of sciences: as 'the history of obsolete knowledge [*connaissances périmées*] and the history of sanctioned knowledge [*connaissances sanctionnées*]', the latter being of contemporary relevance.

Canguilhem gives Bachelard credit for being 'the first to oppose obsolete history to sanctioned history, the history of facts from experiment or scientific conceptualisation appreciated in their relation to the latest scientific values'. He sees this thesis as being utilized by Bachelard to productive effect in many of his own works (EHPS 13; VR 44), but it is clear he recognizes its value for his own approach as well.

Koyré is seen as continuing in this tradition, despite some differences of style and historical period. Deriving from this approach, Canguilhem suggests that 'the history of sciences is not the progress of science in reverse', where obsolete and outmoded approaches are reconstructed. Rather, it is to try to understand why those ideas were held, what made them possible, what made them successful and accepted. In sum, 'to study the history of a theory is to study the history of the doubts of the theorist' (EHPS 14; VR 45).

The fundamental question is: 'how does one do the history of sciences, and how should one do it? This question raises another: *what* is the history of sciences a history *of?*' (EHPS 14; VR 46; see EHPS 9). This is often assumed, rather than explicitly questioned. The externalist approach would describe a set of events, often uncritically assumed to be scientific, 'in terms of their relation to economic and social interests, technological needs and practices, and religious or political ideologies'. Canguilhem suggests that this is 'an enfeebled or rather impoverished version of Marxism' (EHPS 15; VR 47). On the other hand, internalist approaches argue that 'there is no history of sciences unless one places oneself within the scientific endeavour itself in order to analyse the procedures by which it seeks to satisfy the specific norms that allow it to be defined as science rather than as technology or ideology' (EHPS 15; VR 47). Internalist approaches are, of course, criticized by the externalists as idealist.

Canguilhem has doubts about either approach, since he suggests that they both share the same problem. That is that they 'conflate the object of the history of sciences with the object of a science ... a fact in the history of sciences is treated as a scientific fact, according to an epistemological position which consists in privileging theory over empirical data' (EHPS 15; VR 47); 'When we speak of a science of crystals, the relation between science and crystals is not a genitive relation as when we speak of the mother of a kitten.' Rather, it is a 'discourse on the nature of crystals' (EHPS 16). This leads him to the fundamental suggestion of the difference between history of sciences and science: 'the history of sciences is the history of an object – discourse – that *is* a history and *has* a history'. On the other hand, 'science is the science of an object that is *not* a history, that has *no* history' (EHPS 16; VR 25–6). There are, then, at least three levels: the objects which science studies, the scientific discourse about them, and then history of sciences as the study of that scientific discourse. The world does not exist as scientific objects and phenomena, but science constitutes them through its work. The history of sciences is thus a discourse on discourses, and it follows from this that 'the object of the history of sciences has nothing in common with the object of science' (EHPS 17; VR 26).

> The history of sciences is the explicit, theoretical recognition of the fact that the sciences are critical, progressive discourses for determining what aspects of experience must be taken as real. The object of the history of sciences is therefore a non-given, an object whose incompleteness is essential. In no way can the history of sciences be the natural history of a cultural object. All too often, however, it is practised as if it were a form of natural history, conflating science with scientists and scientists with their civil and academic biographies, or else conflating science with its results and results with their contemporary pedagogical enunciation.
>
> (EHPS 17–18; VR 28)

Canguilhem thus makes some important distinctions at a conceptual level, which have a methodological importance. He argues that the history of sciences might need to address different kinds of information and evidence in its approach: 'there are always documents to be classified, instruments and techniques to be described, methods and questions to be interpreted, and concepts to be analysed and criticized'. Yet, while all are important, he contends that the last is the most significant, and it is the only one which 'confers the dignity of history of sciences upon the others'. Concepts are crucial, and without them there is no science. Technological innovations play a role, and so too do institutional academic conditions, accidents, institutions and cross-fertilization of ideas, but the history of sciences is only interested in them to the extent that they shed light on theories, on concepts. His example is striking: 'Descartes needed [David] Ferrier to grind optical glass, but he provided the theory of the curves to be obtained by grinding' (EHPS 19; VR 30; see EHPS 237; VR 112).

It follows from this that 'the history of the results of knowledge [*savoir*] can only be a chronological register' (EHPS 19; VR 30). This means 'the history of sciences concerns an axiological activity, the

search for truth. This axiological activity appears only at the level
of questions, methods and concepts, but nowhere else' (EHPS 19;
VR 30). This does not mean that the historian of science is uninterested in what scientists do, and only in what they think. On the
contrary, this is significant historically as long as the end is kept in
mind, and it is also crucial in the present: 'Only contact with recent
science can give the historian a sense of historical rupture and continuity. Such contact is established, as Gaston Bachelard taught,
through epistemology, so long as it remains vigilant' (EHPS 20; VR
31). Indeed, this approach means that the history of sciences is not
produced as a fixed object through this work, but that it 'is always
in flux, correcting itself constantly' (EHPS 20; VR 31). Drawing on
Cavaillès, Canguilhem says that 'mathematics is a process [*un devenir*]'
(EHPS 20). The history of sciences can thus become a history of
thought, and one of the crucial aspects of this is that the historian
must recognize that 'the same word is not the same concept'. Instead,
the historian 'must reconstitute the synthesis within which the
concept is found, that is both the conceptual context and guiding
intention of experiments and observations' (EHPS 177). While Canguilhem is therefore taking up some of Bachelard's ideas, we should
note, as Althusser stresses, that he tended not to use the famous
term 'break [*coupure*]' systematically.[35]

A recurrent theme in Canguilhem's discussion concerns innovation and influence. There are often examples of people hailed as
the inventors of an experiment or a proof, when that can be found
in many early instances. William Harvey is praised for his work on
circulation of the blood, but earlier thinkers understood at least
parts of this. Aristotle and Galen understood blood in a similar way
to irrigation of a garden, but this crucially will use up water. Hiero-
nymous Fabricius used an experiment similar to Harvey's, but
understood its outcome in a different way – the regulative purpose
of veins was to prevent blood building up in the lower limbs. Harvey's contribution was to show that the heart pumped blood that
weighed more than the body in twenty minutes, and that if an
artery was cut the body would bleed itself dry: 'Thus the idea of a
closed circuit is born' (KL 23/8; see EHPS 227). 'The reality of the
biological concept of circulation presupposes abandoning the con-
venience of the technical concept of irrigation' (KL 23/9).

Canguilhem therefore follows J. T. Clark in speaking of the 'virus
of the precursor', suggesting that 'strictly speaking, if precursors
existed, the history of sciences would lose all meaning, since science
itself would merely appear to have a historical dimension' (EHPS

20–1; VR 49).[36] This type of thinking is a problem, because it 'assumes that concepts, discourses, speculations and experiments can be shifted from one intellectual space to another', but this is to neglect the 'historical aspect' of what is being studied. Many historians have looked for precursors to Darwin in the eighteenth century – a tendency which Foucault criticized in *The Order of Things* (EHPS 21; VR 50).[37] The precursor is therefore 'an artefact, a counterfeit historical object' (EHPS 22; VR 51). He makes a similar critique of precursors in an essay on Darwin and Wallace (EHPS 100). In that essay he also criticizes the idea of convergence, or that a certain 'idea was in the air'. He is clear on that, suggesting that 'this banality … explains and clarifies nothing' (EHPS 100). He continues: 'the air of a time is a pre-scientific concept in the history of sciences, a vague concept of the geography of organisms, uncritically imported into the arsenal of critical literature' (EHPS 100–1). In later work, he praises Limoges's work on Darwin in *La sélection naturelle*, in part because it challenged 'that mainstay of traditional historiography, the concept of *influence*' (IR 19/9). As he concludes the crucial introduction to *Études d'histoire*: 'The history of sciences is not a science, and its object is not a scientific object. To do [*faire*] the history of sciences (in the most operative sense of the term) is one of the functions, and not the easiest, of philosophical epistemology' (EHPS 23; VR 52).

Scientific ideology

In the late 1960s and early 1970s, in the last years of his teaching career, Canguilhem made a significant shift in his thinking. This was to introduce the notion of 'ideology', which he explored through the discussion of what he called 'scientific ideology' or, more specifically, 'medical ideology'. While he would not present this in a full book until 1977 with *Ideology and Rationality*, he dated the shift to a decade before:

> In 1967–68, under the influence of work by Michel Foucault and Louis Althusser, I introduced the concept of scientific ideology into my teaching and some articles and conferences. This was not simply a mark of my interest in and acceptance of the original contributions of those two thinkers to the canons [*déontologie*] of scientific history. It was also a way of refurbishing without rejecting the lessons of a teacher whose books I read but whose lectures I was never able to attend. For whatever liberties my young colleagues may have taken

with the teachings of Gaston Bachelard, their work was inspired by and built on his.

(IR 9/ix)

Some marks of the tension around May 1968 can be found in *Ideology and Rationality*. In its original French edition, it bears the subtitle *Nouvelles études d'histoire et de philosophie des sciences* – New studies of the history and philosophy of science. In English, it was translated as *Ideology and Rationality in the History of the Life Sciences*. The French, clearly, presents the book as a sequel to *Études d'histoire*, first published in 1968. Canguilhem, however, suggests that the introduction of this concept of ideology should not be seen in terms of his own biography, and suggests that the book is not intended to show 'signs of change or evolution in my thinking' compared to *Études d'histoire* (IR 9/ix–x). He begins the preface to the work with the line that 'to err is human, to persist in error is diabolical', and notes that some of the work presented in this book may appear different from some 'methodological axioms' he borrowed and employed at the beginning of his career, 'some forty years ago' (IR 9/ix). Indeed, in his characteristic way, he claims indifference to how his own work is viewed, and even hesitates in presenting these pieces together (IR 9/x). He notes that, having found discontinuity in history, throughout his research in the history of sciences, it 'would be inappropriate to refuse to recognise discontinuity in the history of history'. From this, he derives a somewhat sarcastic motto: 'To each his own discontinuity, his own revolutions in the world of scholarship' (IR 9/x).

While Althusser's political notion of ideology is well known, Canguilhem suggests that the work he is developing also owes much to Foucault's *Archaeology of Knowledge*, which is a book he says provides an 'analysis of scientific ideology' which he has 'found very useful' (IR 10/x). In that work, Foucault distinguishes different '"thresholds of transformation" in the history of knowledge: a threshold of positivity, a threshold of epistemologisation, a threshold of scientificity, and a threshold of formalisation. In my published work I am not sure that I have distinguished as carefully as Michel Foucault might wish among the various thresholds crossed by the disciplines I have studied' (IR 10/x).[38] Canguilhem suggests that the disciplines he has studied have not crossed the threshold of formalization, and disagrees with Foucault about the status of Bernard and Pasteur.[39] He suggests that Bernard's work moved faster in an epistemological register than his empirical results; while Pasteur 'was primarily interested in making a positive contribution to research and not unduly concerned with developing a consistent epistemology' (IR 10/x–xi). Nonetheless, he finds these ideas very helpful in his own analysis, as did Foucault in his.[40] He recognizes his potential limits too. In a characteristically self-effacing comment, he suggests that 'it may be, finally, that my analyses are not sufficiently subtle or rigorous. I leave it to the reader to decide whether this is a question of discretion, sloth, or incapacity' (IR 10/xi).

This again raises for him the question of the relation between epistemology and the history of sciences. Canguilhem suggests that the key thing to note is that there are many more 'manifestos and programmes of research than there are concrete facts [*échantillons*]: Statements of intention are numerous, concrete results meagre' (IR 11/1). There is a history of the history of sciences as a discipline, but epistemology seems to have a shorter history and one unconnected to the history of sciences (IR 11/1). This leads him to some important points about how history works. While a historian is obviously trying to narrate a specific fragment of the past in detail, this cannot be a straightforward sense of its reconstruction. One of his pithy aphorisms sums this up: 'Errors of judgment are accidental, but alteration is the essence of memory' (IR 12/2). Reconstruction in the history of sciences must take account of 'a point that has repeatedly been made about reconstructions in other fields of history –political, diplomatic, military and so on: namely, that, contrary to Leopold van Ranke's dictum, the historian can never claim to represent things as they really were (*wie es eigentlich gewesen*)' (IR 12/2).

Again he refers to Dijksterhuis's comment about the history of sciences as an 'epistemological laboratory', which he had discussed a decade before in *Études d'histoire*. This time, he comments:

> Since elaboration is different from restitution, one may conclude that epistemology's claim to give more than it has received is legitimate. Epistemology shifts the focus of interest from the history of sciences to science as seen in the light of history. To take as one's object of inquiry nothing other than sources, inventions, influences, priorities, simultaneities, and successions is at bottom to fail to distinguish between science and other aspects of culture.
>
> (IR 12–13/3)

This means we must be careful about constructing a sense of the past in relation to the present, as an antecedent for a present sense of things. Nor can we see the past as inevitably leading to the present, where the past is judged as either appropriate or inappropriate by present standards. 'Taken in an absolute sense, the "past of a science" is a popularized [*vulgaire*] concept. The '"past" is a catchall of retrospective inquiry' (IR 13/3). Nonetheless, some historians do try to work in that way, where:

> The history of a science is thus a summary of readings in a specialized library, a repository and conservatory of knowledge produced and expounded from the time of the tablet and papyrus, through parchment and incunabula to that of the magnetic tape. This is, to be sure, an ideal library, a library of the mind, the compilation of a sum of traces. The totality of the past is represented here as an unbroken expanse.
>
> (IR 14/4)

This, however, falls into the trap of simply cataloguing. The history of sciences expects more, especially from epistemology. It requires help in grasping what parts of 'the entire plane of the past' are legitimate or significant moves, and which are not (IR 14/4). He says that this is the conclusion reached in an important paper by Suzanne Bachelard on 'Epistemology and History of Sciences'.[41] Suzanne Bachelard – daughter of Gaston, and translator of Husserl's *Formal and Transcendental Logic* – was a significant philosopher whose most important work was *La conscience de la rationalité*.[42] In this paper on 'Epistemology and History of Sciences', she draws upon Canguilhem's work on the reflex, among many other sources.[43]

Canguilhem provides some examples from the history of plant physiology, and suggests that it 'should now be clear why the past of a present-day science is not the same thing as that science in the past' (IR 15/5). We need to sift, discern and balance. It is important, for example, to recognize that events may be 'theoretically significant or insignificant' in terms of the wider 'trajectory of discourse', even if they initially appear to be more comparable. The past lineage of a present moment might be difficult to discern among potentially comparable paths or trajectories (IR 16/6). This is one of the reasons why epistemologists, philosophers, write about the history of sciences, rather than just scientists, since the questions that arise concern the nature of knowledge and its elaboration. He notes that, of course, the epistemologist will be well aware that there are scientists who have written on the history of sciences 'as respite from their scientific labours' (IR 16/6):

> it is one thing to recognize the existence and value of an epistemological history written by scientists. It is another, however, to argue

that the epistemologist must therefore conclude that he has no special relation to the history of sciences on the grounds that a similar relation can be established between the scientist and the history of sciences, to the great benefit of the latter. Or that the epistemologist must remain an outsider, because while his relation to history may appear similar to that of the scientist, his motivation is fundamentally different.

(IR 17/7)[44]

Scientists naturally draw on their immediate predecessors, who of course drew on theirs. Accordingly, 'science has a natural interest in its own history, even if that interest is not very widespread among scientists' (IR 18/8). But this interest is only part of the 'heuristics of research', and may not extend 'to very remote antecedents, where "remoteness" is to be construed in conceptual rather than chronological terms' (IR 18/8). Of course, the limitations of time mean that scientists often neglect the past for their present concerns. Because the historical is their primary interest, epistemologists are not constrained in the same way that the scientist is, either in terms of time, or in terms of specialism. They can 'range more freely', and their 'breadth of knowledge can compensate for the relative inferiority of their mastery of the latest scientific discoveries and analytic tools' (IR 18–19/8–9). He notes that the concern of epistemology for the history of sciences is a topic of longstanding interest, and suggests that in some ways 'epistemology has always been historical'. As he suggests:

When the theory of knowledge [*connaissance*] ceased to be grounded in an ontology incapable of accounting for the terms of reference employed in the new cosmological systems, it became necessary to examine not just the justifications but the methods of science [*savoir*] itself ... When one thinks of the history of science in terms of the progress of enlightenment, it is difficult to envision the possibility of a history of categories of scientific thought.

(IR 19–20/10)[45]

One of the other essays included in this collection is titled 'What is a Scientific Ideology?', dating from 1969. It explores many of his longstanding themes with this new focus. Canguilhem begins by asking what the history of sciences is a history of? If the answer is the simple 'science', then he stresses that 'one must then specify precisely what criteria make it possible to decide whether or not, at any given time, a particular practice or discipline merits the name science' (IR 33/27). 'Merit' is a key word for him, and designating something as a science is 'a dignity not to be bestowed lightly'. This raises a question of whether the history of sciences should 'exclude or, on the contrary, should it tolerate or even include the history of the banishment of inauthentic knowledge from the realm of authentic science' (IR 33/27). This is an 'epistemological problem concerning the way in which scientific knowledge is historically constituted' (IR 34/28).

Is previous science, once it has been shown to be error, an antiscience? In support, he quotes Bogdan Suchodolski's suggestion that 'the history of science as a history of truth is unrealisable. It is a postulate with an internal contradiction.'[46] How does the notion of antiscience relate to that of ideology (IR 34/28)? He suggests that this is a pressing problem, and that 'the question of ideology arises in connection with the practice of the history of sciences, although many practicing historians have never bothered to ask it' (IR 34/28). How do historians of mathematics deal with mystical properties of numbers and shapes, historians of astronomy with astrology, chemistry with alchemy, psychologists with the past? Often by ignoring it, recognizing it as a forerunner, or as a past source of information, or with embarrassment (IR 34–5/28). Is scientific ideology useful to describe pseudosciences 'whose falsity is revealed solely by the fact that a genuine science has been established to refute their claims' (IR 35/29)?

Of course, the term 'ideology' in its present use derives from the popularization of Marx's use, 'an epistemological concept with a polemical function' (IR 35/29):

> In the meaning that Marx gave to the term, he preserved the idea that ideology inverts the relation between knowledge and the thing known. Ideology, which initially denoted the natural science of man's acquisition of ideas about reality, came to be a term applied to any system of ideas resulting from a situation in which men were prevented from understanding their real relation to the real. Ideology exists wherever attention is diverted from its proper object.
>
> (IR 36/30)

Canguilhem provides a brief discussion of the relation between science and ideology in *The German Ideology*, and suggests that ideology is challenged by science, which tears off its veil. His approach aimed at science, in contrast to the ideological – that is, class-influenced – political and economic work he was criticizing: 'No ideology speaks the truth. Although some are less removed from

reality than others, all are illusory' (IR 36–7/30–1). He notes that
'bourgeois ideologies are reactions which indicate symptoms of
social conflict and class struggle, yet as theories they tend to deny
the concrete problems without which they would not exist' (IR
37/31). Nonetheless, he recognizes that Marx and Engels do not
talk about science as an ideology in *The German Ideology*. Canguilhem's position is a more qualified one, in which he recognizes that
sciences are of course linked to the mode of production, and particularly the way that this relates to or exploits nature. But he stresses
that this does not mean that 'the problems of methods of science
are not autonomous'. As such he draws a distinction between the
sciences and other superstructural issues: 'unlike economic or political theory, science is not thereby subordinated to the dominant
ideology of the ruling class at a particular moment in social relations' (IR 37–8/31). Canguilhem is not entirely clear on the argument
here, but he is certainly distancing himself from the ideas, current
at the time, of bourgeois and proletarian science (see chapter 3).

There are, therefore, differences in approach when it comes to
seeing science itself as an ideology. He suggests that the terms need
to be seen as distinct: 'The history of sciences would need to include
a history of scientific ideologies, explicitly recognized as such' (IR
38/32). One crucial difference is political; another epistemological.

> Scientific ideology, unlike a political class ideology, is not false con-
> sciousness. Nor is it false science. The essence of false science is that
> it never encounters falsehood, never renounces anything, and never
> has to change its language. For a false science, there is no pre-scientific
> state. The discourse of a false science can never be falsified. Hence
> false science has no history.

This is an important distinction, since he wants to examine scientific
ideology, rather than to condemn it as false science. The latter is
not his concern. Scientific ideology is much more interesting to him,
because, crucially, it does have a history: 'A scientific ideology comes
to an end when the place that it occupied in the encyclopaedia of
knowledge is taken over by a discipline that operationally demonstrates the proof of the validity of its "norms of scientificity". At
that point a certain form of non-science is excluded from the domain
of science' (IR 39/33).

He therefore prefers the term 'non-science' to 'anti-science' because
he says the point is that a scientific ideology aspires to scientific
status, in imitation of what already is accepted as science, and therefore a prerequisite for a scientific ideology is the existence of scientific

discourse. This, in turn, requires that there is already a distinction between science and religion (IR 39/33).

One example he provides is atomism. 'Atom' comes from the Greek meaning 'uncuttable' or 'indivisible', and there were various ancient philosophies that took the position that atoms were indeed the building blocks of all things. Yet modern, scientific atomism thought that atoms were not the smallest unit (IR 39–40/33–4). In biology, Maupertuis's views on heredity were displaced by those of Mendel (IR 40–1/34–5):

> Mendel was only interested in hybrids in his repudiation of an age-old tradition of hybrid research. He was neither interested in sexuality nor in the controversy over innate versus acquired traits nor of pre-formations versus epigenesis. He was interested only in verifying *his* hypotheses via the calculation of combinations. Mendel neglected everything that interested those who in reality were not his predecessors at all.
>
> (IR 41/35)

Yet this practice of tracing the transitions is perhaps too close to his previous work. As he says: 'Instructive as it is to study the way in which scientific ideologies disappear, it is even more instructive to study how they appear' (IR 42/36). Canguilhem makes three summing-up points:

a. Scientific ideologies are explanatory systems where the object is hyperbolic, relative to their own borrowed norms of scientificity.
b. In every domain scientific ideology precedes the institution of science. Similarly, every ideology is preceded by a science in an adjunct domain that falls obliquely within the ideology's field of view.
c. Scientific ideology is not to be confused with false sciences, magic, or religion. Like them, it derives its impetus from an unconscious need for direct access to the totality of being, but it is a belief that *squints* at an already instituted science whose prestige it recognises and whose style it seeks to imitate.

(IR 44/38)

Canguilhem says that not all history of sciences will work in this way: 'A history of sciences that views science as a series of articulated facts of truth need not concern itself with ideology. Historians of science who hold this view naturally abandon questions of ideology to the historians of ideas or, worse still, the philosophers' (IR 44/38–9). Yet Canguilhem suggests that, for many others, these are necessary concerns: 'A history of sciences that views science as a progressive process of purification governed by *norms of verification* cannot fail to concern itself with scientific ideology' (IR 44–5/39).

Returning to some themes he has discussed before, he suggests that Bachelard's distinction between obsolete and sanctioned science is valuable, but stresses we must also look at the relation between these types of science. What is seen as the obsolete 'is condemned in the name of truth and objectivity'. But it was not always that way: 'what is now obsolete was once considered objectively true'. What is taken at each moment as truth 'must submit itself to criticism and possible refutation or there is no science' (IR 45/39). For these reasons, he challenges any sense that there can be a strict distinction between ideology and science. Making that division would prevent us from seeing continuity, from seeing traces of ideology in the science that has supplanted it. If we see connections between them, then we will not fall into the trap of 'reducing the history of a science to a featureless landscape, a map without relief' (IR 45/39).

He closes this essay with some general principles for a historian of science, returning to Suchodolski's claims, and agreeing with his suggestion that 'the history of the one truth is a contradiction in terms' (IR 45/39–40):

> The historian of science must work and must present his work on two levels. If he fails to recognize and incorporate scientific ideology into his work, he runs the risk of producing nothing more than ideology himself, by which I mean in this instance a history that is a false consciousness of its object. The closer the historian thinks he comes to his object, the farther he is from the target. His knowledge is false knowledge, because true critical knowledge requires critical perspective; the historian cannot accurately see any object that he does not actively construct. Ideology is mistaken belief in being close to truth. Critical knowledge knows that it stands at a distance from an operationally constructed object.

In these late essays, Canguilhem adds another dimension to his work on the history of sciences, utilizing the insights of Althusser and Foucault in developing his own work.

8

Writings on Medicine

The Normal and the Pathological was Canguilhem's doctoral thesis in medicine, before he worked on broader themes in the history and philosophy of sciences. Nonetheless, he returned to medical issues many times in his career. Chapter 2 has already discussed some later work which reflects on the book's themes. One of the last of these essays was reprinted in _Ideology and Rationality_, the last book Canguilhem published, six years after his retirement. For the first decade or so of his retirement, he continued an active speaking and writing career, and it was only in his final years that he really slowed down. Some of the late texts were commemorations of friends and colleagues, and there is a more reflective and elegiac tone to much of his output. But early retirement years were far from inactive, and it is notable that a number of the essays were concerned with medicine again. Many of those texts are included in _Writings on Medicine_. While this was a posthumous collection, it is one that Canguilhem himself anticipated. Geroulanos and Meyers cite a note titled 'La maladie et le malade, la médecine et le médecin' [Disease and the patient, medicine and the doctor] in which Canguilhem lists a number of pieces of his work. They suggest this is the outline of a planned collection. The note is little more than a scrap of paper, but it lists the following pieces:

- The Idea of Nature in Medical Theory and Practice
- Body and Health
- Is a Pedagogy of Healing Possible?
- Power and Limits of [Rationality in Medicine]
- For Dentists[1]

Another scrap of paper, filed in the same place, suggests what may
be an alternative title: 'Esquisse d'une Critique de la raison médicale
pratique' [Sketch of a critique of practical medical reason].²

The essays that Canguilhem listed are from a range of dates.
'The Idea of Nature' was from 1972, and 'Is a Pedagogy of Healing
Possible?' from 1979.³ 'Body and Health' was largely used to inform
a piece entitled 'Health – Popular Concept and Philosophical
Question', published separately in 1998 and included in its place
in *Writings on Medicine*.⁴ 'The Power and Limits of Rationality in
Medicine', presented at a workshop in Strasbourg on 7 December
1978 on the centenary of Bernard's death, was in Canguilhem's
outline, but it appeared in the fifth edition of *Études d'histoire* in
1983 instead (EHPS 392–411). 'For Dentists' was a 1979 lecture
which was largely reused for other pieces included in the collec-
tion.⁵ The essay 'Diseases', dating from 1989, was added to the
collection. All the essays in the book, together with 'The Power and
Limits of Rationality in Medicine', will be discussed in this chapter.
Some other essays on medical themes, appearing in the collections
Études d'histoire or *Ideology and Rationality* will be discussed along
the way.

The work on medicine also relates to his wider interests in biology
and the history of thought. In particular, he is again concerned with
the idea of experimentation. In a discussion of Bernard, Canguilhem
suggests that 'the historical break [*coupure*] by which modern medicine began' can be situated 'in the idea of experimental medicine
as a declaration of war against Hippocratic medicine' (EHPS 131;
VR 275). Bernard's work on experimental medicine was founded
on a number of principles: 'the principle of the identity of the laws
of health and diseases; the principle of determinism of biological
phenomena; and the principle of the specificity of biological functions, that is, the distinction between the internal milieu and the
external milieu' (EHPS 139; VR 279). The latter distinction is significant in showing that the organism has some autonomy, and to
show the limitations of vitalism. Symptoms of disease might exist
in both the normal and pathological state, and so the concept of
health was significant in understanding them. Bernard's work on
experimental medicine is therefore significant, even though he could
not claim to have invented either the project or the concept. But his
arguments and research refuted objections to this work, and he
came to be identified with it (EHPS 139–40; VR 280–1).

Canguilhem notes in the 1959 essay 'Thérapeutique, Experimentation, Responsibilité' that 'doctors have always experimented, in

the sense that they are always attentive to what they can learn from their actions, when they took the initiative' (EHPS 389). Yet this experimentation on living subjects is hardly conducted in ideal laboratory conditions, is often undertaken in emergency situations, and is always working with individuals. 'Treatment is to undertake an experiment [faire une expérience]' (EHPS 389). But the overall conclusion is clear: 'a medicine concerned with man in the singularity of his living [vivant] can only be a medicine which experiments. One cannot fail to experiment in the diagnosis, in the prognosis, in the treatment' (EHPS 389).

One example is his 1971 discussion of John Brown (1735–88) and his Elements of Medicine (1780). Canguilhem asks why this text, with its 'numbered paragraphs in imitation of [Euclid's] Elements of Geometry' and its 'theory of organic incitability', had so much influence, not in his own England, but in Philadelphia, Italy and Germany (IR 47/41). Canguilhem later adds that Brown had least success among physicians in France, but even there some chemists picked up on his work, though the lack of take-up of his work may be down to there being only a partial translation until 1805 (IR 51/45–6). For Canguilhem, this text is a case of medical ideology.

Canguilhem suggests that Cuvier gives a convincing account, claiming that 'before Brown, the best that physicians could do was to organize their observations of a patient into a history of his disease', and then try to develop a prognosis through analogies. If these analogies could be generalized, then they might be turned into a principle, which might lead to something worth being called a 'medical theory'. Yet these never received lasting agreement, with initial enthusiasm giving way to disillusionment. Brown's theory had some important merits, among them 'extreme simplicity', and led to some positive changes in practices. 'It seemed to reduce the medical art to a small number of formulas', but, while these were 'ingenious', they could not disguise its inadequacies; which endure in the reformulations of his claims by later physicians.[6] Canguilhem reports various accounts of the teaching of medicine. In antiquity, it was claimed that someone could be educated on Galen's theories in six months; under Brown's system, it could be reduced to four weeks (IR 49/43).

Brown's work was a continuation of the mechanistic tradition, and it has been claimed that 'his medical erudition was limited' (IR 50/44): 'In asserting that every organism possesses a finite quantity of incitability, or capacity to be affected by inciting powers or stimuli, he did not trouble himself to provide either justification or evidence but simply set forth a principle by which life was distinguished from inert matter. [Werner] Leibbrand quite aptly insisted that it was a question of "axiomatic force" ' (IR 51/45).[7]

Brown reduced a number of complexities by insisting on equivalence, suggesting that animal and vegetal, agriculture and medicine, nerves and muscles, health and sickness are all the same. Canguilhem gives a number of indicative quotes from Brown, among which are 'the same powers excite all the phenomena of life'[8] and 'everything in nature is the work of a single organ'.[9] He takes from this that 'Brown could describe himself as the Newton of medicine, the first to endow medical theory with the certainty of a true science' (IR 51/45).[10] The lack of adoption in France may be down to the lack of a full translation, or a French insistence on clinical observation, and nosographical classification (IR 51/45–6). But Canguilhem suggests that it was essentially down to 'their general concept of vital phenomena' (IR 52/46).

> Although Broussais saw irritation as the cause of disease [mal] whereas Brown saw stimulation as the remedy, both men shared an intense conviction that normal and pathological organic phenomena are fundamentally identical. This principle, which abolished the distinction between pathology and physiology, was accepted by Magendie, Auguste Comte, and Claude Bernard. For Bernard and others, it became the basis of an ideology, that of medicine's unlimited power, a medical ideology free of every vestige of Hippocratism.
>
> (IR 53/47)[11]

Understandably, given his professional career, Canguilhem finds one thing revealing: 'Regrettably or not, the fact is that today, in order to practice medicine, no one is expected to have the least knowledge of its history' (WM 19/27). This is not just that his career was concentrated on the historical, but that he trained as a doctor in order to be able to do that work. The only thing remaining of Hippocrates is 'the famous oath, the last surviving rite, which has now been emptied of its meaning [sens]' (WM 19/27). He adds that what is worse is that some people 'take it upon themselves to judge Hippocrates as if what comes to us downstream in the course of history also must appear at the source' (WM 19/27). This brings him to another favourite theme, the relation between individual experience and scientific knowledge. In relation to medical power, we must remember 'the patient is a subject, capable of expression' (EHPS 409). It follows from this that 'it is impossible to cancel out the subjectivity of the lived experience of the patient in the objectivity of medical knowledge [*savoir*]' (EHPS 409). Given the interleaved themes of the essays on nature, healing and health, it seems helpful to synthesize some of Canguilhem's thoughts around these questions, along with that of the role of the physician.

Health

Canguilhem wonders, with Epictetus, if we can speak of health before Hippocrates (WM 49/43).[12] Nonetheless, this does not mean that the definition remains static. How should we make sense of health? It is striking that health is often defined as an absence, a silence. He gives various examples. Returning to a quotation he had used in *The Normal and the Pathological*, he notes that Leriche defined it this way: 'Health is life lived in the silence of the organs.'[13] Paul Valéry put it more positively, but in a similar manner: 'health is the state in which necessary functions are achieved imperceptibly or with pleasure'.[14] Both trade on Charles Daremberg's earlier suggestion that 'in health, one does not feel the movement of life; all functions are accomplished in silence'.[15] Later in the essay, he gives the World Health Organization definition: 'Health is a state of complete physical, moral, and social well-being and not merely the absence of disease or infirmity' (WM 60/48).[16] Elsewhere, though, he gives Xavier Bichat's much more active determination: 'life is the ensemble of functions which resist death'.[17]

Canguilhem notes how, in the classic period and in the Enlightenment, the absence of disease was seen as the same as health (WM 51/44), and says that this can be found in claims by Leibniz and Kant, among others. But Kant is intriguing, because he says we can feel well, but not know we are healthy, and his definition makes 'health an object outside the field of knowledge. We can put Kant's statement more strongly: there is no science of health. Let us accept this for the moment. Health is not a scientific concept; it is a popular [*vulgaire*] concept. Which is not to say that it is trivial, but simply common, within everyone's reach' (WM 52/44–5).[18] He contends that Descartes's 'conception of health is especially important, since he is the inventor of the mechanist conception of organic functions' (WM 52–3/45). In a 31 March 1649 letter, Descartes had said: 'Health is the greatest among all those of our goods which concern our body, it is however the one which we least reflect upon and savour. The knowledge of truth is like the health of the soul: once a man possesses it, he thinks no more of it.'[19]

Canguilhem thinks this is significant, but raises a question: 'How is it that no one ever thought of reversing this comparison, that no one ever asked whether health were the truth of the body? Truth is not only a logical value specific to the exercise of judgment. There is another sense of truth – for which we do not need to turn to Heidegger' (WM 53/45).[20] In Littré's dictionary, Canguilhem finds the definition that truth is the 'quality by which things appear such as they are' (WM 53/45).[21] He suggests that this is 'a thesis waiting for an author' (WM 54/45), but finds support in Nietzsche. Even in the light of many commentators, he suggests, it is not easy 'to determine the sense and scope of Nietzsche's many texts relating to disease [*maladie*] and health' (WM 54/46).

As might be expected, Canguilhem rejects mechanism, again: 'Health, the body's truth, does not arise out of an explanation of theorems. There is no health for a mechanism ... For a machine, the operative state is not health, and disorder is not a disease ... There is no disease of the machine, just as there is no death of the machine' (WM 57–8/47). This leads to him to a clear formulation of the main focus of all his work, biological or medical:

> The living body is thus the singular being [*existant*] whose health expresses the quality of the forces [*pouvoirs*] that constitute it: it must live with the tasks imposed on it, and it must live exposed to an environment that it does not initially choose. The living human body is the totality of the powers of a being [*pouvoirs d'un existant*] that has the capacity to evaluate and represent to itself these powers, their exercise and their limits.
>
> (WM 58–9/48)

It is for these reasons that he claims that 'the body is at once a given and a product. Its health is at once a state and an order' (WM 59/48).

Broadening beyond the individual, Canguilhem recognizes that issues of health concern not just the singular body, but the social body of the collective, the population. It is here that Canguilhem begins to bring in themes about hygiene or public health, traditionally understood as different branches of medicine:

> The hygienist endeavours to govern [*régir*] a population – individuals are not his business. 'Public health' is a contestable term – 'salubrity' would be more appropriate. Very often, what is public, publicized, is disease. The patient calls for help, draws attention; he is dependent. The healthy man who adapts silently to his tasks, who lives the truth of his existence in the relative freedom of his choices, exists in

> a society that ignores him. Health is not only life lived in the silence
> of the organs – it is also life lived in the discretion of social relations.
>
> (WM 62/49)

As Canguilhem, like Foucault, explores, some of this work on hygiene means that health's 'existential meaning has been occulted by the demands of accounting ... *health* was starting to lose its meaning as truth and to receive a meaning as facticity. It was becoming the object of a calculation. Ever since, we have known this as the health check-up [*le bilan de santé*]' (WM 60/48). In summary, what we call health and its relation to the body has become 'lived assurance – in the double sense of insurance against risk and the audacity to run this risk' (WM 61/49).

The developments of public hygiene, of antibiotics and of chemotherapy transformed the idea of the cure: 'The statistical calculation of therapeutic performances introduced into the understanding of the cure an objective measure of its reality' (WM 80/57; see IR 69/65). But, in terms of conceiving of the relation between the body and society, Canguilhem underlines that 'organic integrity was a metaphor for social integration before the metaphor was inverted' (WM 74/55). Yet there are naturally collective issues about individual health as well. When asked, if we say we are well, then it stops a conversation; if we say we are unwell, then it leads to further interrogation, and to whether 'I am registered with social security'. As Canguilhem puts it: 'Interest in individual organic failure is eventually transformed into interest in the budgetary deficit of an institution' (WM 62–3/50).

While Canguilhem does have an interest in collective health, he remains most focused on the idea of health as the 'truth of the body', and suggests that it should be understood as 'its very constitution or its authenticity of existence' (WM 63/50): 'The definition of health that includes the link of organic life to pleasure and pain tested as such surreptitiously introduces the concept of a *subjective body* into the definition of a state that medical discourse thinks it can describe in the third person' (WM 64/50).

This relates to comments he makes about the importance of the individual, as the medical subject, rather than as an object of medical science. He thinks that philosophy is crucial here, as it provides the basis for seeing that health is a concept rather than just a medical phenomenon. Health should be 'treated as a concept on which popular experience confers the meaning of a permission to live and act according to the well-being of the body' (WM 65–6/51). This is also important in understanding the body, which itself should be grasped as a collective. He notes that, in health today, the concentration of specialisms means that the body is understood as divided: 'The diffusion of an ideology of medical specialization often results in the body being lived as if it were a battery of organs' (WM 66/51).

In an earlier essay Canguilhem had discussed the idea of people being their own doctor, either through modification of behaviour or trusting in nature, and he tracks the notion to the titles of some books from the late seventeenth century, even though the practice predates this (WM 26/30). The idea of naturalism in medicine is significant, but this work is 'divided between two intentions or two motivations: a sincere reaction of compensation in times of crisis in therapeutics and an astute utilization of the distress of the patient for the sale of some snake oil [orviétan], even in print form' (WM 26/30–1). This notion of healing through nature has ended up confined to popular literature, because of 'the conjunction of anatomical pathology and new techniques of clinical exploration, such as percussion and auscultation' (WM 28/31; see IR 58/54). Auscultation is 'listening to the internal sounds of the body, of respiration, blood flow, abdominal movement'; percussion, 'tapping the body to listen for the resonance of masses in the body'.[22] The contrast of medical specialism to 'nature's spontaneous silence' is significant here: 'nature only speaks if one interrogates it well' (WM 28/31).

Yet, despite his critique, he does not go as far as Illich. Canguilhem rarely makes direct reference to him, but suggests that 'the same man who militated for a society without schools called for an insurrection against what he named "the expropriation of health"' (WM 67/51). The reference is to Illich's book Medical Nemesis: The Expropriation of Health.[23] Canguilhem is being quite critical, suggesting that 'this defence and illustration of "natural, private health" in order to discredit "scientifically controlled health", has taken many forms, including the most ridiculous' (WM 67/51–2). He is certainly disputing that his own position inevitably leads to the rejection of 'the scleroses that are considered consequences of eruditely controlled behaviour ... the supposedly oppressive tutelage of medicine and, beyond that, of the sciences that medicine applies' (WM 67–8/52).[24]

A decade before these lines, he had explicitly related 'antimedicine' to the work being done in 'antipsychiatry'. He suggested that both were to some degree intellectually dishonest, in that they exploit 'the initial advantage gained from arguments that assume as true what they need to prove' (WM 94–5/64). However, returning to the title he floated as a model for this collection, he raises some concerns of his own: 'But it seems the time has come for a *Critique of Practical Medical Reason* that would explicitly recognize, within the ordeal of healing, the necessity of collaboration between experimental knowledge and the propulsive non-knowledge of this a priori opposition to the law of degradation, for which "health" expresses a success that always becomes suspect again' (WM 98/65).

In the essay on health, he pays tribute to Merleau-Ponty, making reference to his book *The Visible and the Invisible*, and his courses *The Union of the Body and Soul* and *Nature*.[25] In the first, there is a note: 'The Cartesian idea of the human body as human *non-closed* – open inasmuch as governed by thought – is perhaps the most profound idea of the union of the soul and the body' (WM 65/50–1).[26] Canguilhem returns to Merleau-Ponty at the end of the essay, as his way of justifying 'having made health a philosophical question'. He finds the answer in *The Visible and the Invisible*: 'philosophy is the set of questions wherein he who questions is himself implicated in the question' (WM 68/52).[27]

Healing and the role of the doctor

If natural or self-healing is largely discredited, healing remains what the patient expects from the doctor, though not always what they receive. Canguilhem notes that 'there is thus a discrepancy between the patient's hope regarding the power that he attributes to the doctor on the grounds of the latter's knowledge and the doctor's recognition of the limits of his own efficacy' (WM 69/53). This is the principal reason, Canguilhem claims, that 'of all the objects specific to medical thought, healing is the one that doctors have considered the least' (WM 69/53). There are complications here, because healing seems subjective, in the terms of the patient's assessment of the process; but it is also objective in the general sense of 'a statistical survey of its results', which can only be evaluated beyond the single patient, and the failure of this in one patient does not invalidate the process, just as success cannot necessarily be put down to the process. There may be all sorts of other complications in the mix (WM 70/53–4). A patient may feel that healing is what they need from the doctor; while the doctor, on behalf of medicine, gives them 'the best-studied, best-tested, and most-used treatment currently available' (WM 71/54). A doctor, then, is not the same as a healer: the doctor is licensed by their knowledge, a healer by their successes (WM 71/54).

Canguilhem suggests that the doctors who addressed the question of healing were mainly psychoanalysts, or ones who used that process to examine their practice critically (WM 71–2/54):

> It is known that according to its etymology, 'to heal [*guérir*]' is to protect, to defend, to arm – quasi-militarily – against aggression or sedition. The image of the organism offered here is the image of a city menaced by an external or internal enemy. To heal [*guérir*] is to guard, to shelter [*garder, garer*]. This was the idea well before certain contemporary physiological concepts, such as aggression, stress, and defence, entered the domain of medicine and its ideologies. The likening of healing to an offensive–defensive reaction is so profound and originary that it has penetrated even the concept of disease, considered as a hostile reaction to some invasion or disorder.
>
> (WM 73/55)

Returning to one of the figures so influential in his work in biology, Canguilhem suggests that Goldstein's work on the organism has not received sufficient attention beyond Merleau-Ponty and his influence. 'Perhaps this is because Goldstein himself presented his thesis as an epistemology of biology, rather than as a philosophy of therapeutics' (WM 91/62). Following Goldstein, he suggests that 'a healthy organism comes together with the surrounding world in such a way as to be able [*pouvoir*] to realize all of its capacities. The pathological state is the reduction of the initial latitude for intervening in the milieu' (WM 91/62).

Canguilhem is clear that the doctor–patient relationship is never a 'simple, instrumental relation', and is always 'mediated by something foreign to its space of intelligibility' (WM 15/25). He suggests that we must not 'conceive the doctor–patient relation as the relation of a competent technician to a machine that is out of order [*dérangé*]', even though the formal training of doctors 'prepares them very poorly to admit that healing is not exclusively determined by interventions of a physical or physiological order' (WM 85/60). The doctor's ambition is, however, clear: 'To procure, for the sick person [*malade*], by efficacious interventions, an amelioration or a restitution that he would not know how to obtain by his own means' (WM 16/25). That may imply, though, that 'the sick organism, is, vis-à-vis the doctor and for him, nothing more than an object that is passive and obedient to external manipulations and solicitations … For an inert body, an active medicine' (WM 16/26). This links to claims made in the essay on the question of healing: 'In popular terms, to heal is to recover a compromised or lost good, namely, health' (WM 81/58), but there is a subjective element, where some patients will accept less than they might be owed, and others will not see themselves as healed even when what they needed has been done (WM 82–3/59).

Yet equally, and conversely, recognizing the limited power of medicine goes alongside a recognition that the living body has 'a spontaneous capacity of conserving its structure and regulating its functions', and this means that the body's 'own capacities of defence' must be trusted, 'at least temporarily', and that this 'is a hypothetical imperative for both prudence and skill. For a dynamic body, an expectant medicine' (WM 16–17/26).[28] This forces a recognition of the limits of medicine: 'Not all treated patients heal. Some patients heal without a doctor' (WM 17/26). Canguilhem makes reference to Dagognet, who shows that modern medicine often works against the body's own self-defence reactions, actively trying to moderate or even suppress them (WM 19–20/27).[29] Medicine may therefore need to work against the body for its benefit. This leads him to the claim that 'it is sometimes too little to say that nature's remedy is worse than the disease – it is itself disease and harm, it is itself evil [le mal, il est le mal lui-même]' (WM 21/28). Examples of this might be 'allergy and anaphylaxis' (WM 21/28).

Contemporary medicine may not be Hippocratic, because it doubts natural organic defences, but it is not anti-Hippocratic: 'Nature's capacity to cure is not denied by a treatment that governs it by integrating it – it is located in its proper place, or more exactly, it is comprised within the treatment's limits … Today, ignorance would consist in not asking of nature what is not its own. The medical art is the dialectic of nature' (WM 22/29).

> From the moment when medicine founds its diagnosis no longer on the examination of spontaneous symptoms, but on the examination of provoked signs, the patient's and the doctor's respective relations to nature are turned upside down. Because he cannot make out the difference between signs and symptoms on his own, the sick person is led to believe that any conduct that is regulated exclusively by symptoms is natural. For his part, the doctor now knows that he cannot accept all that nature says and the way in which it says it without first using his art to force nature to express itself; and for this reason, he is thus led to defy nature – not only in what it says, but also in what it does.
>
> (WM 28–9/31–2)

The first part of this claim is dependent on Foucault's work in *Birth of the Clinic*. Physiology therefore supports some natural approaches, through its ongoing research into self-regulation and stabilization, and work on the 'therapeutics of infectious diseases' has helped us to understand the way that organisms have 'an innate capacity of antitoxic defence' (WM 29–30/32). Yet physiology is active, rather than passive. We should not wait for nature to reveal itself, but, rather, should mobilize its resources: 'To act is to activate, as much to reveal as to cure' (WM 31/33). It follows that we can continue 'to speak of "nature" when designating the initial fact of existence of self-regulating living systems whose dynamics are inscribed in a genetic code' (WM 31/33).[30] It follows that a pedagogy of healing 'should seek to obtain the subject's recognition of the fact that no technique, no institution, whether currently or in the future, will assure the integrity of his powers of relating to men and things' (WM 98/65).

Diseases

Late in life, Canguilhem wrote a short essay on the notion of diseases.[31] He begins with quoting Diderot's remark in his *Notes on Painting* that 'nature does nothing that is not correct. Every form, whether beautiful or ugly, has its cause; and of all the extant beings, there is not a single one that is not just as it should be'.[32] He continues to say that we could imagine an *Essays of Medicine* which began in a similar way: 'Nature does nothing arbitrarily. Disease, like health, has its reason, and among all the living beings, there is not one whose state is not as it must be' (WM 33/34). Like Diderot, he is clearly working with Leibniz's principle of sufficient reason. Yet he quickly cautions that 'this kind of prologue could not concern all populations at all times' (WM 33/34).

As we might expect, he then begins a discussion of how a history, and to some extent a geography, applies to notions of disease: 'Over centuries and in many places, illness was either a possession by some "malignant [*malin*]" being against which only a miracle worker [*thaumaturge*] could triumph or a punishment inflicted by a supernatural power on the deviant or the impure' (WM 33/34). We can find this in the Old Testament as much as in the Far East, with the obvious example being leprosy. Indeed, he notes that what we call Ancient Greek medicine really only applies since Hippocrates. Hippocrates' thought was 'the moment when diseases came to be treated as bodily disorders about which it is possible to construct a communicable discourse concerning their symptoms, their supposed causes, their probable course, and the behaviour that must be observed if one is to correct the disorder that all these indicate' (WM 34/35). It is a parallel development to the emergence of what might be called science and philosophy – even if the medicine of this time is not ' "scientific" in the modern sense of the term' (WM 34–6/35). As he clarifies: 'Contemporary medicine is founded, with an efficacy we cannot but appreciate, on the progressive disassociation of disease and the sick person, seeking to characterise the sick person by the disease, rather than identify a disease on the basis of the bundle of symptoms spontaneously presented by the patient. Disease points us to medicine rather than to evil [*mal*]' (WM 35/35).

> Plague, cancer, shingles, leukemia, asthma, and diabetes are also species of organic disorder that are felt by the living as a harm or an evil [*comme un mal*]. Disease is the risk of the living as such – risk as much for the animal or the vegetal as it is for the human. Yet in this last case, and by contrast with the risk that is born from a resolution to act, the risk that is born with birth is quite often inevitable. Suffering, the restriction of chosen or required habitual activity, organic deterioration, and mental decline are all constitutive of a state of harm [*mal*], but are not by themselves specific attributes of what the physician of today identifies as disease, even at the moment when he endeavours to put an end to this harm [*mal*] or to attenuate it.
> (WM 35–6/35)

Today, medicine is increasingly becoming 'a science of diseases', but this has the effect sometimes of putting the patient 'between brackets', even though they are in some sense the focus of medical concern (WM 36/36).

A whole set of transformations, inventions and sometimes crises in medicine has led to increased knowledge of the 'structure of the pathogenic agent', and led physicians 'to change the target of reparative intervention'. Diseases have been increasingly and 'successively localized in the organism, the organ, the tissue, the cell, the gene, the enzyme' – a set of locations – and the work to identify them has also been done in a series of locations: 'the autopsy room, then in laboratories where physical examinations (optical, electrical, radiological, scannographic, echographic), then chemical or biochemical examinations are conducted' (WM 36–7/36).

Medicine and biology have become more closely linked, which has led to distinctions between hereditary diseases, inborn (*congenital*) diseases, and occasional diseases which are 'results of the relations of the individual to the ecological milieu as well as to the social group', which may include 'individual accidents, such as pneumonia, or collective ones, such as the flu or typhus, that is, diseases known as infectious' (WM 37/36). For such diseases, there has been much work on 'causes (microbes, bacilli, viruses) … vectors (the flea on the rat for the plague, the *Aedes aegypti* mosquito for yellow fever)', as well as a need 'to attend to the reasons of their geographical distribution or to the form of social relations proper to affected populations' (WM 38/36–7). There are a whole range of complicating factors, from the question of microbes resistant to vaccines, to the increasingly artificial human environments we live in, to 'innate errors of metabolism' or 'hereditary biological anomalies' which mean that some populations are more disposed to poisoning or disease, while others are resistant to them. For this reason, there are diseases seen by the enzymologist – that is, someone who studies biochemical reactions – which might not be of concern to the physician (WM 39/37).

The relation between living conditions and disease is also important. On the one hand, the separation of the knowledge of a disease from the living conditions of a patient is a product of 'the colonization of medicine by the general and applied sciences since the early nineteenth century' (WM 40/37). But it is 'also an effect of the interest (in every sense of the term) of industrial society since that period in the health of working populations, or, as some put it, in the human component of productive forces' (WM 40/37–8). Hygienists encouraged political authorities to take more seriously 'the monitoring and amelioration of living conditions'; and a parallel development was found in the shift from hospices or shelters, where the sick were housed but not treated, to a hospital understood as a 'curing machine [*machine à guérir*]' – a term Canguilhem attributes to Jacques Tenon, but which is common to the period (WM 40/38):[33] 'Treatment of diseases in hospitals under a regulated social structure contributed to their deindividualization at the same time that the progressively more artificial analysis of their conditions of appearance separated their reality from their initial clinical representation' (WM 40–1/38).

There is also a corresponding shift in the notion of a doctor: 'The doctor–therapist concerned with all aspects of illness and currently known as "general practitioner [*généraliste*]" witnessed declining prestige and authority in favour of specialist physicians, engineers who have taken apart the organism like machinery' (WM 41/38). These are no longer doctors in the standard, lay sense, with much more attention paid to statistical information, semiology (the science of signs) and aetiology (the study of causes) (WM 41/38). The statistical change occurred at the same time as the 'anatomical–clinical revolution' of the nineteenth century in Western Europe: 'In short, one cannot deny the existence of a social, hence political, component in the invention of theoretical practices that are currently effective in the knowledge of diseases' (WM 41/38).

This question can be seen in at least two ways. One would be 'the investigation of causes of a change in medical knowledge and direction', but the other would be 'causalities of a sociological order in the appearance and course of diseases themselves' (WM 42/38). This has been analysed by trade unionists talking of 'capitalism's diseases – which amount to seeing in disease the organic index of class relations in capitalist societies'; or of 'diseases of misery or poverty [misère]' that result from deprivation, such as 'vitamin deficiencies resulting from malnutrition'; or of earlier concerns in the eighteenth and nineteenth centuries about hygiene, population density and occupational health (WM 42/38–9): 'Nevertheless, whatever importance in the increase of pathological situations we should accord to patterns of life and their relations to working conditions … it is improper to confuse the social genesis of diseases with the diseases themselves' (WM 42–3/39). However, there are some cases 'in which the census and evaluation of disease factors can take into account the social status of patients and their representations of it … The fact of living with a disease … is one that we must consider among the components of the disease itself' (WM 43–4/39). This is on the 'nebulous border between somatic medicine and psychosomatic medicine … haunted by psychoanalysis, because what is in question is as much the unconscious as the techniques devoted to making it speak in order that we may know how to respond' (WM 44/39). There is, however, a resistance:

> We must recognise that currently methods of identifying diseases and therapy are due to the successes of immunology, rather than to thaumaturgies inspired by psychosociology. Immunology is a biochemical discipline based on medical experiment [expérience] … The conceptual revolution concerning diseases was the identification of what came to be called the immune system, that is, a totalizing structure of responses to aggressive antigens by way of the production of specific antibodies. The collaboration of clinic and laboratory in immunological research is perhaps still fragile, but it has introduced the reference to biological individuality in the representation of disease.
> (WM 44–5/40)

This has challenged the divide between medical and scientific understandings of disease, which was sometimes sharp in the nineteenth century, but the alliance has led to 'the recent extinction of smallpox as a result of the preventive vaccination measures derived from Pasteurian bacteriology' and a 'common hope to find someday, through molecular biology, an effective response to diseases that are nowadays burdened with phantasms of distress: cancer and AIDS' (WM 45/40). The existence of disease is both 'a universal basic fact and a singular existential test of man'. It requires us to make an 'interrogation of the precariousness of organic structures', and he suggests that medicine until now has not made a sufficiently 'convincing response' to this challenge (WM 47/41).

He draws on letters that Freud, suffering from cancer, addressed to Lou Andreas-Salomé, and concludes: 'Diseases are instruments of life by which the living (when it is man we are talking about) see themselves forced to avow their mortality' (WM 48/41). He had made some related claims a decade before, suggesting that 'man is open to disease not by being condemned to it or destined for it, but because of his simple presence in the world' (WM 89/62). He adds that, 'in this regard, health is not at all an economic exigency to be asserted within a legislative framework; it is a spontaneous unity of the conditions for the exercise of life' (WM 89/62).

Bacteriology and medical theory

The point about bacteriology links to a longstanding interest. In a 1975 essay on the topic, Canguilhem suggests that 'nothing in the history of science is more instructive than putting in synchronic relation a successful practice with theories that were hostile to it, because they were incapable of establishing a deterministic frame which would legitimate the practice'. His first example is vaccination: the use of the cowpox virus to inoculate humans against smallpox. This was much safer than inoculation with smallpox itself, and became such an effective treatment that the smallpox disease was eventually eradicated. But Canguilhem notes that 'none of the medical systems of the time was even remotely capable of explaining the statistically measurable success of the treatment or to account for certain failures' (IR 55/51).

Crucially, in order for there to be an advance in medicine capable of making sense of this development, and making the most of its potential, the change did not come from medicine itself. It came from chemistry, and Canguilhem notes that the path from eighteenth-century medicine to nineteenth 'was not without detours. One need not be a Hegelian to admit that in medicine, too, history rarely follows a straight line.' Medicine remained 'a symptomatology and nosology explicitly based on the classification of the naturalists', revived old doctrines and borrowed indiscriminately from other sciences, and objected to innovation on philosophical grounds. The borrowing from other sciences was not new: Francis Bacon had wanted it to learn from chemistry, and Descartes from mechanics, but 'experimental medicine' took a long time to establish itself – 'The phrase remained a signifier in search of a signified.' Equally, 'therapeutics alternated between sceptical eclecticism and obstinate dogmatism, but without a foundation other than empiricism'. In sum, Canguilhem suggests, 'medicine could not accomplish its project: it remained an empty discourse about practices often not very different from magic' (IR 56–7/52–3).

There are three main reasons things changed in the nineteenth century, according to Canguilhem. First, there was what Foucault calls 'the birth of the clinic', with hospital reforms and developments in practice – percussion, auscultation and 'systematic efforts to relate observed symptoms to anatomico-pathological data'. Second was 'a rational attitude to therapeutic scepticism'. Third, there were developments in physiology – separating itself from classical anatomy, and focusing on the tissue, and later the cell; and looking to 'physics and chemistry for examples as well as tools'. From all this, 'a new model of medicine was elaborated' – 'new diseases were identified and distinguished'; old medications were abandoned; 'rival medical theories cast discredit on one another'. A new medicine emerged, based on experiment, 'knowledge without system', all leading, in aspiration at least, to 'effective therapies, whose use could be guided by critical awareness of their limitations' (IR 58/54–5). But this model was still an ideology, and 'if the goal of the program was eventually achieved, it was reached by a detour and by routes quite different from those envisioned by the program's authors' (IR 59/55).

To explore these issues, Canguilhem discusses various medical thinkers, including Brown, Broussais and Bernard (IR 59–63/55–8). The specific details of their approaches need not detain us here, but Canguilhem is interested in how the debate between their approaches was characterized. Bernard accused Broussais of creating a 'system', not a theory – which Canguilhem suggests is perhaps better understood as a 'medical ideology' (IR 61–2/57):

By scientific ideology – the notion of which is still controversial for many – I mean a certain type of discourse that parallels the development of a science and that, under the pressure of pragmatic needs, goes beyond what has actually been proved by research. So that as a discursive construction it is, in relation to a science which it will qualify as an ideology, both presumptuous and misplaced. Presumptuous because it believes that the end has been reached when it stands at the beginning. Misplaced because the promises of the ideology, when achieved by science, are *different* and in *a different area*.

(IR 62/57–8)

Canguilhem suggests that Broussais's promise had already begun to be delivered by François Magendie (IR 62/58). But it did not take the same form: 'In short, Magendie's experimental medicine differed from Broussais's physiological medicine in a triple displacement: in place, the laboratory rather than the hospital; in experimental object, animals rather than … men; in an external or internal modifier, instead of Galenic principles it used extracts isolated by pharmaceutical chemistry, for example replacing opium with morphine and quinquina with quinine.' The second displacement was treated with 'the greatest incomprehension and criticism', largely because it abolished 'the distance between' animals and men. He was accused of experimenting on humans, which was a charge he denied. 'But if administering unproven drugs is experimentation … then Magendie did experiment on humans, patients in hospitals, which he considered a vast laboratory where the population allowed the constitution of groups and their comparison' (IR 63/58–9). Magendie's body of work is remembered today as 'surpassed but not repudiated', though 'there was a striking gap between his therapeutics and his physiology', still dominated by a 'Hippocratic practice of watchful expectation' (IR 63/59). He denied contagion in the case of yellow fever and cholera; amended French legislation on quarantines, a practice 'invented in the fourteenth century by the cities of Venice and Marseilles'; and 'understood nothing of the physiological mechanism of anesthesia and vehemently opposed its use in surgery' (IR 63–4/60).

Bernard described Magendie as 'the ragpicker of physiology … merely the initiator of experimentation. Today it is a discipline that has to be created, a method.'[34] As the discussion of his work in chapters 3 and 4 showed, Bernard put a strong emphasis on experimentation, planning a late but uncompleted work, 'Treatise on Experiment in the Medical Sciences'. For him, theory and progress go together, and Canguilhem suggests that not enough attention has been given to this aspect of his work: 'Experimental medicine is progressive, he argued, because it elaborates theories and because those theories are themselves progressive, that is, open' (IR 65/61). Canguilhem provides two key quotes from Bernard: 'An experimentalist never outlives his work; he is always at the level of progress', and 'With theories there are no more scientific *revolutions*. Science grows gradually and steadily.'[35] Canguilhem comments that, when the 'two new concepts of determinism and action' were added to this understanding – and he stresses that the two were dependent on each other – 'you have the four components of a medical ideology that clearly mirrored the progressive ideology of mid-nineteenth-century European industrial society' (IR 65/61).

This view of theory and revolution has been rightly criticized in the light of Kuhn's structure of scientific revolutions and Bachelard on the epistemological break (IR 65/61): 'In short, the concept of a theory without revolution, which Bernard took to be the solid basis of his methodology, was perhaps no more than a sign of internal limitations in his own medical theory. Experimental medicine, active and triumphant, which Bernard proposed as a definitive model, was the medicine of an industrial society' (IR 66/62). Nonetheless, Canguilhem is still critical of his limitations. Bernard 'was never able to appreciate fully that science requires not only that the scientist abandon ideas invalidated by facts but also that he give up the personalized style of research', which was the hallmark of his work (IR 66/62). Equally, Canguilhem suggests that, 'convinced of the identity of the normal and pathological, Bernard was never able to take a sincere interest in cellular pathology or germ pathology' (IR 67/63).

In the year after Bernard's death, Pasteur read a paper titled 'The Theory of Germs and its Applications to Medicine and Surgery' to the Academy of Medicine in Paris. His theory was one that 'already promised health and longer life to millions of men and animals' and 'also spelled the end of all the medical theories of the nineteenth century' (IR 67/63): 'Although the concept of resistance was later shown to play an important part in the relation of microbe to organism, certain uses of concepts from Bernardian physiology undeniably constituted a true obstacle to the therapeutic advances made toward the end of the century by students of Pasteur and Koch' (IR 68/64).

The notion of an obstacle was explored in chapter 7. Here Canguilhem notes Rudolf Virchow's role in 'transporting cellular theory from the laboratory where it was born to the pathology department

of the hospital clinic' (IR 69/64). But its practical role was confined
to surgery, and in the treatment of tumours through 'microscopic
anatomical pathology' (IR 69/64–5).

Canguilhem gives various other examples. Chemotherapy was
an innovation in the history of medicine as 'a therapeutic technique
that was both effective and unrelated to any medical theory' (IR
69/65). Paul Ehrlich's basic formula is summarized by Canguilhem
as asking 'through what chemical compounds with special affinity
for certain infectious agents or cells could one act directly on the
cause rather than on the symptoms of disease, in imitation of the
anti-toxins in various serums' (IR 69–70/65). Treatment of diphtheria
was developed not through a vaccine but 'injection of serum taken
from a convalescent patient, provided one has a convalescent patient,
that is, a survivor of the disease' (IR 70/65–6). Canguilhem does
not want to get into questions of priority here, but he notes that
Ehrlich 'dreamed that chemistry could provide a power far beyond
that of nature' (IR 70/66). Ehrlich made some discoveries himself,
but his real success was 'in those that would ultimately be discovered in pursuit of his fundamental hypothesis: that the affinities of
chemical stains could be used as a systematic technique for developing artificial antigens' (IR 70–1/66). There are complications, of
course, including resistance in bacteria to the drugs, and that organisms sometimes 'defend themselves … against their chemical guard-
ians', leading to 'combined treatment regimes'. But these developments
were made possible by the progress of Ehrlich's work (IR 71/67).
Research led to advances in chemical engineering, and the development of things that were 'impossible and inconceivable in the time
of Magendie' (IR 71/67). The context and the science go together
as 'chemotherapy could not have existed without a certain level of
scientific and industrial society' (IR 72/68), but it also depended
on serotherapy: 'This historical condition in turn depended on other
historical conditions, which too many historians of medicine have
been inclined to view as mere accidents incidental to various kinds
of technological activity' (IR 72/68).

Canguilhem equally recognizes that Pasteur's 'laboratory work
was directly affected by what was going on in the world of technology'. This was not because of practical technical issues raised by
'industrialists, artisans and animal breeders' but because of the
nature of his encounter with chemistry (IR 73/69). Canguilhem is
explicit that he is building on the work of Dagognet on Pasteur at
this point.[36] Pasteur's major contribution was that he 'linked germ,
fermentation, and disease in a unified theoretical framework' (IR

74/70). At this point, Canguilhem returns to the three key displacements from the eighteenth to nineteenth century – 'of place, from
hospital to the laboratory; of object, from human to animal; of means,
from Galenic preparations to chemical compounds'. But these required
a fourth change: 'Pasteur did not find the solution to the pathological problems of the living being in the living. He found it by shifting
his attention to crystals, to those geometric embodiments of pure
mineral substance' (IR 74/70). Instead of 'treating living beings as
if they were inert', he made progress by 'distinguishing living things
from inert substances in terms of their most general structural properties' (IR 74/70).

While some immediately saw the merits of the work, there was
also resistance. Canguilhem takes the example of Léopold Ollier:
'This was not the first time that a man who benefited from the
practical consequences of a theoretical discovery failed to grasp
either the origins or the meaning [sens] of that discovery' (IR 75/71).
Surgery benefitted from 'aseptic and antiseptic practices' as well as
anaesthesia – 'the surgeons were the first to benefit from Pasteur's
research' (IR 75/71). Lister, following Pasteur, recommended use
of phenolic acid for sterilization of wounds 'only twenty years after
the brilliant and pitiful [Ignaz] Semmelweis (died 1865) was forced
to resign from the obstetrical clinic in Vienna for having required
his students to wash their hands' (IR 75/71).

But the relation of medicine to other sciences should be approached
cautiously: 'It was not until the end of the century, however, that
medicine obtained new techniques and substances capable of acting
on a disease that were neither symptomatic nor imaginary' (IR 75/71).
Yet it was not simply that medicine drew from other disciplines:

> It should be laid down as a general principle in the history of science
> that in a given period – and certainly since the seventeenth century –
> discord and rivalry within the scientific community can never totally
> impede communication. For one thing, it is impossible not to be
> affected in some ways by what one rejects. For another, even where
> exchange is impossible, everyone goes to the same market for sup-
> plies. In the nineteenth century the market was primarily one of
> instruments and raw materials. Despite differences of goals, moreo-
> ver, the sciences generally delineate a common field of exploration.
>   (IR 75–6/71)

As he notes, 'hence it is impossible to imagine how a body of medical
knowledge [savoir] such as bacteriology could have been produced
in the nineteenth century without owing something to the contagion

of medical theories that it helped to relegate to the realm of ideology' (IR 76/72).Through the development of practices such as autop-
sies, pathologists searching for internal lesions, and the 'microbial
theory of infectious disease', there were changes in the practice of
examination, and also in the unravelling of previous 'ambiguous
pathological concepts such as fever and inflammation' (IR 76/72).

* * *

These late essays on medicine cycle back to some of his earliest
concerns, while showing the benefit of his many years of teaching
and research. They link to his wide interests in biology, and the
history of the life sciences, as well as his longstanding reflections
on the philosophy of history and epistemology. In bringing together
many of them in the *Writings on Medicine* collection, his editors have
helped to show the thematic unity of his concerns in this regard.
He stresses the importance of understanding health as a philosophical concept as much as a medical phenomenon. He insists on the
irreducible nature of the organism, and the distinctive approach
needed in thinking about it biologically. In reflecting on the health
of the individual body as well as the collective one, there is a strong
ethical and political sensibility to his work. Yet while he is critical
of some of the supposedly unproblematic assumptions of medicine,
he is concerned with trying to improve, rather than abandon, it.
His insistence on the patient's experience is a crucial part of this.
Equally, as shown by his analysis of bacteriology, he continues to
insist on the worth of taking a historical approach to these issues.

9

Legacies

Canguilhem's theoretical propositions about the history of science, its relation to epistemology, and questions of ideology are significant. They are, of course, put to work in his own studies of the normal and the pathological, of biology, of regulation, evolution, psychology and cognition, and medicine discussed in previous chapters. Yet they have implications beyond his own specific interests. What is the legacy of his work?

'Textes et documents philosophiques' and the 'Galien' series

Given Canguilhem's lifelong commitment to teaching and to pedagogical strategy in philosophy, it is not surprising that there are some traces of his organizational work remaining. In particular, Canguilhem edited a book series for Hachette entitled 'Textes et documents philosophiques'.[1] Each volume included a short selection of excerpts from classical texts, arranged thematically. As well as being designed to be resources for teaching, they insist on the importance of tracing any topic historically.

Canguilhem edited the series' first volume on *Besoins et tendances* [Needs and tendencies],[2] and collaborated with Suzanne Bachelard and others on the two-volume *Introduction à l'histoire des sciences*.[3] Limoges notes that the first volume developed directly out of Canguilhem's teaching on the topic, and that he had made such compilations of short texts available to his students from the 1930s.[4]

The history of sciences volumes developed from his teaching at the Institut d'histoire des sciences in the late 1960s. Other series contributors included Deleuze on *Instincts et institutions* and Dagognet on *Sciences de la vie et de la culture*.[5] Fourteen volumes were published in total, all in the 1950s except the history of sciences volume, and one on *Technique et technologie* by Guillerme, which appeared in the early 1970s.[6]

Besoins et tendances begins with a set of texts about elementary biological needs – hunger and thirst, sleep, generation and society. It moves to a study of social needs in humans; the limits of human resistance; sleep, forgetting and refusal; philosophical work on tendencies and desire; foundation and classification of needs; values and technologies – and closes with a brief summary passage from Plato's *Protagoras*. The authors included are those who might be expected, given the topic and Canguilhem's interests in philosophy, natural history and biology elsewhere – Alain, Bergson, Bichat, Freud, Goldstein, Lamarck, Leibniz, Nietzsche, Spinoza, Vidal de la Blache and many others, but also some more surprising ones such as T. E. Lawrence, Marx, Marcel Proust, François Rabelais and the explorer and novelist Antoine de Saint-Exupéry. Some of his contemporaries, such as Hyppolite and Sartre, are also included. The only words directly by Canguilhem are a brief 'Présentation de la collection', which also serves as an introduction to the series as a whole and is reprinted in some subsequent volumes (OC IV, 438–42).[7] There, Canguilhem notes that the purpose of the series was to serve the needs of students and their work, rather than to be used by teachers for students (OC IV, 438). He also acknowledges the artificial nature of this enterprise, and makes the point that obviously none of the texts included were written to be excerpted and arranged in this way (OC IV, 440–1).

More revealing, perhaps, is the volume on the history of sciences. In his brief 'Avant-Propos' to this text, Canguilhem notes the work of students and researchers at the Institut d'histoire des sciences which went into the volume.[8] He suggests that the history of sciences is still a neglected study in France, not part of undergraduate education, and taught only in a few Faculties of Letters and Human Sciences, rather than the natural sciences, and practised 'only in some research centres or institutes, almost all in Paris'.[9] Hoping to improve that situation, the two-volume text was designed to make available a sampling of material. Canguilhem and his colleagues note that it cannot be exhaustive, but that the selection, 'to our mind at least, includes the necessary tools [*instruments*] to go further and elsewhere'.[10] The first volume is organized with an initial section on the History of the History of Sciences, and then Bibliographies. The sciences treated are Logic, Mathematics, Astronomy, Physics, Chemistry, Biology, Sociology, (Political) Economy, Psychology and Linguistics. The arrangement of material is more progressive in mathematics than in physics, and more organized for astronomy than for biology – a product, Canguilhem suggests, of the nature of the subject matter.[11] Designed for a student audience, these books perhaps do not hold much interest for an analysis of Canguilhem's wider research career. But they underline the importance of pedagogy to his career, as an early example of a teaching source that has now become more common, and shows the role he played as a facilitator.

Canguilhem also edited the 'Galien' series with Presses Universitaires de France. A number of crucial works appeared in this series, named after the Greek physician Galen, though nearly all the authors had some kind of relation with Canguilhem. It included both the first and second edition of Foucault's *Naissance de la clinique*; Lantéri-Laura, *Histoire de la phrénologie*; Limoges, *La sélection naturelle*; Ulmann, *De la gymnastique aux sports modernes*; three works by Dagognet; and one by Jean-Charles Sournia.[12] Canguilhem's own *Le normal et le pathologique* was reprinted in the series, and it also included translations of Paracelsus' *Oeuvres médicales* and the Darwin – Gaston de Saporta correspondence.[13] The final volume, published the year after he retired, was Evelyne Aziza-Shuster's *Le médecin de soi-même*.[14] Many of these are works which Canguilhem cites in his own writings, already mentioned in previous chapters. What is striking, aside from the medical theme that structures the series, is that they are all contributions to the *history* of medicine. Many of these authors went on to significant careers, and some of these books were their first works, based on a thesis, often under Canguilhem's direction. Aziza-Shuster's *Le médecin de soi-même*, for example, was a topic Canguilhem suggested and it was examined by Dagognet and Foucault. If there is something of a narrow grouping to the series, based on personal connections, it is also an indication of his support of his colleagues and students in their careers. He regularly used his institutional positions to support that work, creating, among other things, a receptive atmosphere for work loosely labelled as poststructuralism. The influence of Canguilhem on some of his seminar colleagues was also discussed in relation to the collaborative volume *Du développement à l'évolution* in chapter 6.

Canguilhem, Foucault and biology

Rabinow suggests that Canguilhem's legacy is to be found in the 'rejuvenation of the history of biology in France', and that his work as a whole can be seen as an examination of how a number of specific concepts needed to be resituated within medical fields of knowledge.[15] He claims Canguilhem helped shift it 'from a panegyric to progress to a precise laying out of the movement of concepts'. Rabinow suggests, though, that Canguilhem's work merely hints at some of the 'immense social and political implications' of his research, and that Foucault, especially in the 1960s, was concerned with developing and making explicit this immanent potential.[16] Rabinow argues that Foucault's work in that period was 'to an important extent centered on the history of what we now call biology'.[17]

Indeed, Canguilhem apparently told Foucault after the thesis defence: 'Now it's time for you to get to work.'[18] Foucault published *Birth of the Clinic* not long afterwards, in the 'Galien' series. However, the result of this injunction can perhaps best be seen in the book *Les mots et les choses*, with its examination of the transition from natural history to biology. Equally, Rabinow argues that *The Archaeology of Knowledge*, 'whatever its limitations … is a gold mine of insights and suggestions for the study of the history of scientific objects', as it outlines 'a comprehensive program for the study of conceptual objects, subjects and strategies'.[19] As Canguilhem himself notes, 'In the *History of Madness* and *Birth of the Clinic*, Michel Foucault brilliantly demonstrated how the methods of botany served as a model for nineteenth-century physicians in developing their nosologies' (EHPS 340; VR 306–7). In a copy of his Introduction to Charles Kayser's collection on physiology, Canguilhem wrote a note to Foucault, calling his essay the 'unfaithful disciple' of Foucault's histories of madness and the clinic.[20]

Of course, these themes continue through Foucault's later work as well. In the late 1960s, Foucault seriously discussed a project on the idea of heredity, some of which was developed in a course on sexuality at the University of Vincennes in 1969.[21] He suggested this as a future project in his candidacy to the Collège de France, and discussed it in his inaugural lecture there.[22] Heredity is a theme which brings together a range of Foucault's interests – madness, medicine, crime, sexuality and biology. The project was dropped as punishment became his key focus in the first part of the 1970s, but it continued in his discussion of how evolutionary biology became important in psychiatry, in inherited characteristics, the notion of monstrosity and criminal profiling. Some of his lectures discuss the politics of medicine beyond the hospital, taking in themes about public health and epidemics. His notion of biopolitics first emerges here: 'the body is a biopolitical reality, medicine is a biopolitical strategy'.[23] In different places, Foucault discusses smallpox, the plague and cholera. Broader biological themes can be found in his analysis of race, of population, and in his work on sexuality. One of the initially planned volumes of the *History of Sexuality* was on populations and races, and from the indications we have of this project, one of its themes would have been the dual formation of a biology of reproduction and a medicine of sex.[24] Another planned volume in the series was on the pervert, and this seems to have been where Foucault intended to develop themes from his lectures around abnormality and monstrosity. His work on hermaphrodites, of which the edited memoir of the case of Herculine Barbin is the only published part, looks at the relation between biological sex and social questions. Although Foucault completed none of these volumes, and his *History of Sexuality* project took a very different course, even his analyses of pagan and Christian antiquity some-times discuss medicine, the question of life, and what we would now call biology.[25]

François Delaporte

One of the figures in contemporary French thought whose work bears the closest relation to that of Canguilhem is François Delaporte. Delaporte edited *A Vital Rationalist*, the collection of Canguilhem's work in English, but it does neither of them justice as a mark of scholarship. Delaporte's thesis was conducted with Foucault's support, on *Les Questions de la végétalité au XVIIIe siècle*, and was published as *Le second règne de la nature* [*Nature's Second Kingdom*], with a preface by Canguilhem.[26] Delaporte has gone on to write detailed studies of the 1832 cholera outbreak in Paris, yellow fever, Chagas disease and myopathies, as well as a popular introduction to epidemics.[27] In recent years, he has continued work in this vein, with a focus on facial expression and musculature, in relation to physiology and the emotions; on face transplants; and on parasitic disease.[28] In the preface to his study of the cholera outbreak, Delaporte acknowledges his dual debt: 'Michel Foucault some years ago sug-gested that I work on this subject and was kind enough to read the manuscript in draft. Georges Canguilhem gave me invaluable advice.'[29] Equally, in his study of yellow fever, he states that his work is indebted to Canguilhem, whom he thanks for 'his encouragement and advice'. He adds that he has 'paid close attention to his views on methodology, as well as those of Michel Foucault'.[30]

In his work on cholera, Delaporte makes the claim that 'disease does not exist. It is therefore illusory to think that one can "develop beliefs" about it or "respond" to it. What does exist is not disease but practices.'[31] This should not be misunderstood as suggesting that what we call diseases are fictions, or the phenomena that make them up are imaginary. What it means is that, as Rabinow puts it, 'we can know the world only through our elaboration of concepts'. The practices studied are medical, as is the science or knowledge of them.[32] The relation between words, concepts and practices is significant here. His study of Chagas disease is a case in point. Today, this is the name for American trypanosomiasis, a tropical parasitic disease. We might point to a number of moments in this story. Clearly, something existed in terms of the symptoms long observed in people. The insects which spread the parasite are now known to be triatominae, known as conenose or kissing bugs or other, local, names. Carlos Chagas described the disease in 1909, but for Delaporte this is not the crucial moment. The symptoms Chagas described, the protozoa he attributed to them, and the vector of transmission he established related to a wide range of conditions, not all of which we would describe as the disease today, without a specific and systematic articulation. As Delaporte notes, the title of his book 'masks a fundamental ambiguity', and he says the point of his history is to 'dispel it'.[33] The key for Delaporte was scientific work in the 1930s, although the role of Salvador Mazza in the Argentinian Gran Chaco province, who supposedly described sufficient cases with clinical precision, is subject to a demythologization here. The role of Cecilio Romaña in describing the swelling of the area around the eye socket, known as Romaña's sign, is also significant.[34]

There is obviously not the space here for a detailed discussion of Delaporte's valuable work, but it is clear that his writings on both biology and medicine take up themes and approaches from Canguilhem, even if they develop them in a quite different set of analyses. Canguilhem did, however, write prefaces to two of Delaporte's studies, and so his analysis of his protégé is worth a little attention. On his study of vegetable life, Canguilhem suggests that Delaporte is able to provide a discussion of 'the break [*rupture*] in continuity between the work of eighteenth-century naturalists

and early nineteenth-century biologists, a break that marks the
beginning of progress that has continued without interruption to
the present day'.[35] More generally, he suggests that 'the author dis-
plays great erudition and yet never lapses into pedantry'.[36]

Canguilhem's preface to Delaporte's study of yellow fever is also
important:

> Nothing seems more simple today, in speaking of an epidemic disease,
> than to distinguish between the focus and specific agent, between
> the mode of transmission and diffusion, between the distribution of
> disease across populated areas and the composition of the living
> milieu. Yet the appropriate concepts of germ, vehicle, and intermedi-
> ate host had to be laboriously elaborated by means of observation,
> analogy, experimentation, and refutation (some would say falsifica-
> tion). In the case of yellow fever, great subtlety was required in ana-
> lysing the conditions resulting in infection. Properly speaking there
> is no transport unless the object transported is the same at the point
> of arrival as at the point of departure. If an alteration takes place
> along the way, what occurs is no longer transmission in the strict
> sense. The concept of *vector* was elaborated to combine two concepts
> that had proved useful in understanding other infections, that of
> vehicle and that of host. The *Culex* mosquito takes up the germ,
> harbours it during a period of incubation, and injects a product that
> it has in a sense fashioned.[37]

Canguilhem also notes the importance of broader contextualizing
factors in Delaporte's work. This comes through strongly in his
history of Chagas disease, which begins with a discussion of how
colonial explorers, and the naturalists who followed them, described
the bugs and the disease. It proceeds to look at the coffee plantation
industry, and the need to protect the health of the workers: 'Pro-
moting public health was an instrument of exploitation.'[38] As Brazil
modernized, this programme of public health became a political
and economic priority. Institutions were established, medical pro-
fessionals trained, and foreign scientists came to the country. This
instituted a nosopolitics – a term Delaporte takes from Foucault to
describe a politics of health and disease.[39] Canguilhem finds similar
issues in the study of yellow fever: 'The history of this conceptual elucidation, whose validity was demonstrated by its practical
consequences for treatment and prevention, is also the history of
various individuals, actors whose functions, work, and responsibilities
involved them in the history – properly speaking, the political history
– of the exploitation of the globe, of colonization, of international
trade.'[40]

The political context is more explicit than in most of Canguilhem's own studies, though he would certainly acknowledge that there are social and political aspects to much of what he examines. But he stresses that Delaporte is exemplary as a historian too, one clearly in his own image, partly because he is a historian of science who takes epistemology seriously. This, Canguilhem, suggests, allows him to make an 'equitable determination of the respective contributions of each of the leading figures in this lengthy investigation'. In a more general register:

> He has been able to avoid such errors as confusing a precursor with a founder, a word with a concept, a transport with a cycle of transmission. He has been able to discern, in this arduous and sometimes nasty controversy, an important biological advance, namely, the 'complete redefinition of the alliances among living things'. These alliances are sometimes deadly, just as in human societies. It is certainly true that the elucidation of yellow fever's mode of transmission altered the figure of Death. It became possible to trace on a map of the earth the boundaries within which Death has wings.[41]

Even though the specific nature of his inquiry is significant, like his own studies, Canguilhem sees in here the general precepts of how to conduct a historical investigation.

* * *

The focus here on Delaporte is not intended to diminish Canguilhem's influence elsewhere, but to look at a lesser-known figure who is working very much along the lines Canguilhem established. Canguilhem's influence was, indeed, extensive. As Foucault famously put it:

> Take away Canguilhem, and you will no longer understand very much about a whole series of discussions that took place among French Marxists; nor will you grasp what is specific about sociologists such as Bourdieu, [Robert] Castel, [Jean-Claude] Passeron, what makes them so distinctive in the field of sociology; you will miss a whole aspect of the theoretical work done by psychoanalysts and, in particular, by the Lacanians. Furthermore, in the whole debate of ideas that preceded or followed the movement of 1968, it is easy to find the place of those who were shaped directly or indirectly by Canguilhem.[42]

Whether this is something of an exaggeration is debatable, but his work has been utilized, outside of his own interests, in many fields.[43] As Foucault added elsewhere, part of Canguilhem's influence came through his university positions, as teacher and administrator, but his students were diverse: 'Many of his students were neither Marxists nor Freudians nor structuralists.'[44] Other names that might be added would include Deleuze, Roudinesco and Gilbert Simondon. Bruno Latour's early studies of laboratory life and science owe something to Canguilhem, even though he is far more critical of the idea of science and its truths.[45] Although Canguilhem said nothing about gender or feminism, even here his work has been used productively.[46] In a related way, Ian Hacking has shown how Canguilhem's ideas can be put in dialogue with Donna Haraway, even though her early work on biology, conducted in France, makes no reference to him.[47] Canguilhem has also had a small influence in contemporary medical geography.[48] Most recently, Bernard Stiegler's *Automatic Society* begins with a discussion of milieu, indebted to Canguilhem, in which the importance of external influences is analysed, especially as part of a toxic work culture.[49] Catherine Malabou has also returned to Canguilhem's discussion of psychology in her own work on intelligence.[50]

Braunstein has rightly suggested that, in France, 'Canguilhem is no longer just considered as the disciple of Bachelard and the master of Michel Foucault, even if that is important in itself.'[51] In the Anglophone world, we still lag some way behind that, forgetting Canguilhem's own comments about the problem of the precursor. Nonetheless, the lineage was partly created by Canguilhem himself, and partly through the work of Lecourt. Foucault helped to cement it when he made an apparently strict distinction between a 'philosophy of experience, of meaning [*sens*], of the subject' and one of 'knowledge [*savoir*], rationality and of the concept'. While Sartre and Merleau-Ponty were in the first camp, Foucault put Canguilhem, along with Bachelard, Koyré and Cavaillès, in the second.[52] He suggested that the closest comparison outside of France was the Frankfurt School.[53] Yet the distinction, for Canguilhem at least, is more blurred than might appear.[54] There is a philosophy of life at the heart of Canguilhem's project, as much as the knowledge of it. As Foucault puts it later in this essay, 'through an elucidation of knowledge [*savoir*] about life and of the concepts that articulate that knowledge, Canguilhem wishes to determine the situation of the *concept in life* ... Forming concepts is a way of living and not of killing life.'[55] Indeed, Foucault concludes that Canguilhem proposed a 'philosophy of error, the concept of living, as another way of approaching the notion of life'.[56]

While Bachelard's influence has been discussed, and the links to Foucault outlined, the focus of this study has been Canguilhem's own fascinating projects. There are many things to take from his work, from his approaches to questions of history and epistemology to the broader contributions to medicine and biology. For Canguilhem, life is irreducible to mechanist understandings, biology is a distinct science that should not use other sciences as its foundation, and organisms are inextricable from their milieu. The politics of his work was strong at the outset of his career, and comes through in different places, especially in the work of the 1940s with his role in the resistance to Nazism and the Vichy regime. Yet, even much later in his career, it provides a powerful set of ideas and influences. This book has attempted to show how Canguilhem deserves to be read and thought with, as a significant thinker in his own right. In one of his historical studies, Canguilhem comments that 'a book is not read because it exists. It only exists as a book, as a deposit of meaning [*dépôt de sens*], because it continues to be read' (EHPS 128). His work continues to be read, almost despite himself.

One of Canguilhem's very last lectures, given at the ENS on 10 March 1990, at the award of the Prix Jean Cavaillès, asked 'What is a philosopher in France today?'[57] He examined the popular perception of philosophy, and the professionalization of the discipline with professors, noting that, until recently, few philosophers held institutional positions. He also examined some moments of which he had first-hand experience – from Sartre's 'Existentialism is a Humanism' to the rise of Althusser, Foucault, Lévi-Strauss and Lacan in the 1960s. He drew on a range of recent examples to think about the role of a philosopher in a technocratic society. It was an appropriately valedictory topic, over which he ranged widely. Canguilhem never ceased to reflect critically on the question of philosophy, a topic he always approached historically.

Timeline

A more detailed chronology can be found as Camille Limoges, 'Critical Bibliography', VR, 385–454. A biography and complete bibliography is promised in the forthcoming *Oeuvres complètes*, vol. VI.

4 June 1904	Born in Castelnaudary
1924	Enters École Normale Supérieure
Nov. 1927 – April 1929	Military service
1929–40	Teaching in *lycées* and medical training
1935	*Le fascisme et les paysans* (anonymous pamphlet)
1939	*Traité de logique et de morale* (with Camille Planet)
22 June 1940	French armistice with Germany
	Quits teaching position and joins the resistance
1941	Joins University of Strasbourg philosophy department
1943	*The Normal and the Pathological* published
Feb. 1944	Jean Cavaillès executed
Feb.–May 1947	Three lectures to Jean Wahl's Collège philosophique
1948	Becomes Inspector General of Philosophy
1950	*The Normal and the Pathological* second edition

1952	*Besoins et tendances* – first volume of Hachette series – published
1952	*Knowledge of Life* published
1953	UNESCO report on teaching of philosophy
1955	*La formation du concept de réflexe aux XVIIe et XVIIIe siècles* published
1955	Succeeds Bachelard to chair in history and philosophy of the sciences at the Sorbonne; and as Director of the Institut d'histoire des sciences et des techniques of the University of Paris
18 Dec. 1956	'What is Psychology?' lecture to Collège philosophique, Sorbonne
1958–60	Two years of seminar 'Du développement à l'évolution au XIXe siècle'
19 April 1960	Writes report on Foucault's *Histoire de la folie* thesis
9 Feb. 1962	'Monstrosity and the Monstrous', lecture in Brussels
1962	*Du développement à l'évolution au XIXe siècle* published
1963–72	Edits 'Galien' series in the history of medicine for Presses universitaires de France
1965	*Knowledge of Life* revised edition with additional essay
1966	*The Normal and the Pathological* revised edition with new essays
1968	*Études d'histoire et de philosophie des sciences* published
1970–1	*Introduction à l'histoire des sciences* (two volumes) in Hachette series
1971	Retires from Sorbonne and the Institut d'histoire des sciences et des techniques
1977	*Ideology and Rationality* published
20 Feb. 1980	'The Brain and Thought' lecture at the Sorbonne

1983	George Sarton Medal of the History of Science Society *Études d'histoire et de philosophie des sciences* revised edition with additional essay
1987	Médaille d'or of the Centre National de la Recherche Scientifique
9–11 Jan. 1988	Michel Foucault, Philosophe conference
7 May 1988	'La santé, concept vulgaire et question philosophique', Strasbourg, originally published as a small booklet and reprinted in *Writings on Medicine*
10 Mar. 1990	'Qu'est-ce qu'un philosophe en France aujourd'hui?' – lecture at the awarding of the Jean Cavaillès prize, ENS
23 Nov. 1991	Michel Foucault conference, Sainte-Anne
11 Sept. 1995	Dies in Marly-le-Roi
2002	*Writings on Medicine* published posthumously
2011–	Publication of the first volume of his *Oeuvres complètes* – six volumes planned, three published to date

Notes

All works referenced are by Canguilhem unless otherwise indicated.

1 Foundations

1 Gaston Bachelard, *Études*, Paris: Vrin, 1970; Bachelard, *L'engagement rationaliste*, Paris: Presses Universitaires de France, 1972.
2 *Études d'histoire*, consciously or not, reverses the subtitle of a Festschrift for Bachelard to which Canguilhem contributed: G. Bouligand, G. Canguilhem, P. Costabel et al., *Hommage à G. Bachelard: Études de philosophie et d'histoire des sciences*, Paris: Presses Universitaires de France, 1957. Canguilhem's essay 'Sur une épistémologie concordataire', 3–12, is reprinted in OC IV, 729–39.
3 David Macey, 'Obituary: Georges Canguilhem – Philosopher Who Put Science in its Place', *The Guardian*, 9 October 1995, 11.
4 See Jean-François Sirinelli, *Génération intellectuelle: Khâgneux et normaliens dans l'entre-deux-guerres*, Paris: Fayard, 1988, 464–5, based on an interview with Canguilhem on 7 May 1981.
5 Pierre Bourdieu, 'Georges Canguilhem: An Obituary Notice', trans. Graham Burchell, *Economy and Society* 27 (2–3), 1998, 190–2, 191.
6 On Aron, see Jean-Claude Chamboredon (ed.), *Raymond Aron, la philosophie de l'histoire et les sciences sociales*, Paris: Éditions ENS, 1999, which includes a memoir by Canguilhem, 'La

problématique de la philosophie de l'histoire au début des années 30', 9–23.

7 Sirinelli, *Génération intellectuelle*, 647–59. Canguilhem recalls that, while he met Hyppolite at this time, he only got to know him when they were colleagues at Strasbourg twenty years later: 'Jean Hyppolite (1907–1968)', *Revue de métaphysique et de morale* 74 (2), 1969, 129–36, 129; see 'De la science et de la contre-science', in *Hommage à Jean Hyppolite*, Paris: Presses Universitaires de France, 1971, 173–80.

8 Bourdieu, 'Georges Canguilhem', 191.

9 'La théorie de l'ordre et du progrès chez A. Comte', *Diplôme d'études supérieures*, 1926, CAPHÉS GC 6.1. On Bouglé, see Mike Gane, *French Social Theory*, London: Sage, 2003, 125–26.

10 Sirinelli, *Génération intellectuelle*, 49, 155; Camille Limoges in VR 388.

11 Sirinelli, *Génération intellectuelle*, ch. XVII; Louise Ferté, Aurore Jacquard and Patrice Vermeren (eds.), *La formation de Georges Canguilhem: Un entre-deux-guerres philosophique*, Paris: Hermann, 2013, pt II.

12 For a recollection of his impact, see Jacques Piquemal, 'G. Canguilhem, professeur de Terminale (1937–1938): Un essai de témoignage', *Revue de métaphysique et de morale* 90 (1), 1985, 63–83.

13 *Le fascisme et les paysans*, Comité de Vigilance des Intellectuels Antifascistes, 1935 (reprinted in OC I, 535–93).

14 Colin Gordon, 'Canguilhem: Life, Health and Death', *Economy and Society* 27 (2–3), 1998, 182–9, 186.

15 G. Canguilhem and C. Planet, *Traité de logique et de morale*, Marseille: F. Robert et fils, 1939 (reprinted in OC I, 633–924).

16 Jean-François Braunstein, 'Canguilhem avant Canguilhem', *Revue d'histoire des sciences* 53 (1), 2000, 9–26; Xavier Roth, *Georges Canguilhem et l'unité de l'expérience: Juger et agir 1926–1939*, Paris: Vrin, 2013; and Ferté et al. (eds.), *La formation de Georges Canguilhem*.

17 Camille Limoges, 'Critical Bibliography', in VR 403; David Macey, 'Georges Canguilhem, 1904–1995', *Radical Philosophy* 75, 1996, 56.

18 Sirinelli, *Génération intellectuelle*, 464; David Macey, 'The Honour of Georges Canguilhem', *Economy and Society* 27 (2), 1998, 171–81, 176; see EGC 122–3.

19 For his initial teaching, see OC IV, 55–63; for the reading list on logic and mathematics he inherited from Cavaillès, see OC IV, 1151–9.

20 *L'enseignement de la philosophie: Enquête internationale de l'Unesco*, Paris: UNESCO, 1953; *The Teaching of Philosophy: An International Enquiry of UNESCO*, Paris: UNESCO, 1953. Canguilhem's contributions are reprinted as OC IV, 529–600.
21 Dominique Lecourt, 'Georges Canguilhem, le philosophe', in Jean-François Braunstein (ed.), *Canguilhem: Histoire des sciences et politique du vivant*, Paris: Presses Universitaires de France, 2007, 27–43, 28. See Samuel Talcott, 'The Education of Philosophy: From Canguilhem and *The Teaching of Philosophy* to Foucault's *Discipline and Punish*', *Philosophy Today* 61 (3), 2017, 503–21.
22 Bourdieu, 'Georges Canguilhem', 191; Dominique Lecourt, *Georges Canguilhem*, Paris: Presses Universitaires de France, 2008, 116.
23 The proceedings are published as *Michel Foucault, philosophe: Recontre internationale Paris 9, 10, 11 janvier 1988*, Paris: Seuil, 1989; and *Penser la folie: Essais sur Michel Foucault*, Paris: Galilée, 1992.
24 Michel Foucault, *Dits et écrits*, ed. Daniel Defert and François Ewald, 4 vols., Paris: Gallimard, 1994, vol. IV, 775; Foucault, *Essential Works*, ed. Paul Rabinow and James Faubion, 3 vols., London: Allen Lane, 1997–2000, vol. II, 476. On Nietzsche, see also 'De la science et de la contre-science', especially 177–80.
25 Michel Fichant, 'George Canguilhem et l'idée de la philosophe', in *Georges Canguilhem: Philosophe, historien des sciences – Actes du colloque (6-7-8 décembre 1990)*, Paris: Albin Michel, 1993, 37–48, 48 n. 4. On the relationship, see Barbara Stiegler, 'De Canguilhem à Nietzsche', in Guillaume le Blanc (ed.), *Lectures de Canguilhem: Le normal et le pathologique*, Paris: ENS Editions, 2000, 85–101.
26 For a detailed discussion of this tradition, but beginning much earlier, see Cristina Chimisso, *Writing the History of the Mind: Philosophy and Science in France, 1900 to 1960s*, London: Routledge, 2016 [2008]. Canguilhem is the last major figure treated.
27 Jean-François Braunstein, 'Présentation', in Braunstein (ed.), *Canguilhem*, 10.
28 See 'Descartes à travers mes âges', CAPHÉS GC 28.1, folder 5: 'Descartes interrompu …, 1948'. In GC 28.1.1, subfolder 1, there are letters from the publisher, including one on 20 March 1947 which thanks him for the signed contract. On 27 March he is thanked for providing the beginning of the book, with a view to receiving the rest by July and publishing in October. The folder contains handwritten and some typed versions of several

parts of the book. There is no correspondence concerning the failure to deliver the book. See also Braunstein and Schwarz, OC I, 490 n. 1.

29 'L'analogue et le singulier dans la science, Sorbonne, 1957–1958', CAPHÉS GC 13.3.1, 3–5.

30 Letter 4 December 1990, in Michel Deguy, 'Allocution de cloture', in *Georges Canguilhem: Philosophe*, 323–30, 324.

31 The course can be found as 'Les normes et le normal, 1942–43', CAPHÉS GC 11.2.2.

32 'Normal et pathologique, norme et normal', CAPHÉS GC 15.1.1.

33 'Histoire de la tératologie depuis Étienne Geoffroy Saint-Hilaire, 1961–62', CAPHÉS GC 12.2.8; 'La monstruosité et le monstrueux, Bruxelles, 9 février 1962', CAPHÉS GC 26.1.6.

34 See, for example, 'Science et idéologie dans la constitution de la psychologie, 1967–68 (2e semestre)', CAPHÉS GC 17.3.1; 'L'idéologie médicale au XIXe siècle, 1969–70', CAPHÉS GC 17.3.4; and 'L'idéologie médicale au XIXe siècle, neurologie et psychiatrie, 1970–71', CAPHÉS GC 17.3.5.

35 For a few exceptions, see OC IV, 1097–1111 (dialogue with Badiou); OC IV 1121–38 (discussions with Badiou, Hyppolite, Foucault, Dina Dreyfus and Paul Ricoeur).

36 Bourdieu, 'Georges Canguilhem', 191–2; see Macey, 'The Honour of Georges Canguilhem', 172–3.

37 Macey, 'Georges Canguilhem', 56.

38 Bourdieu, 'Georges Canguilhem', 190.

39 J. Guillerme, 'La détermination numérique des actions médicamenteuses', in Georges Canguilhem (ed.), *La mathématisation des doctrines informes: Colloque tenu à L'institut d'histoire des sciences de l'Université de Paris*, Paris: Hermann, 1972, 67.

40 Guillerme, 'La détermination numérique des actions médicamenteuses', 69.

41 Macey, 'Georges Canguilhem', 56. See Jean-Paul Aron, *Les modernes*, Paris: Gallimard, 1984, 11; Lecourt, *Georges Canguilhem*, 113–21.

42 'Ouverture', in *Penser la folie*, 39–42. See Foucault's 1965 letter to Canguilhem, in Didier Eribon, *Michel Foucault*, trans. Betsy Wing, London: Faber, 1991, 103.

43 Daniel Defert, 'Chronologie', in Foucault, *Dits et écrits*, vol. I, 22.

44 Foucault, *Dits et écrits*, vol. I, 167; Foucault, *History of Madness*, trans. Jonathan Murphy and Jean Khalfa, London: Routledge, 2006, xxxv–xxxvi.

45 Eribon, *Michel Foucault*, 102; based on an interview with Canguilhem, and clarified in a letter from Eribon to Canguilhem,
8 March 1988, CAPHÉS GC 33.7.7.
46 'Sur l'Histoire de la folie en tant qu'événement', *Le Débat* 41,
1986, 37–40, 38. For a longer discussion of this process and
relationship, see Stuart Elden, *The Early Foucault*, Cambridge:
Polity, forthcoming.
47 Michel Foucault, *Histoire de la folie à l'âge classique*, Paris: Gallimard, 1972 [1961]; Foucault, *History of Madness*; Foucault,
Naissance de la clinique: Une archéologie du regard médical, Paris:
Presses Universitaires de France, 1963, 2nd revised edn, 1972;
Foucault, *The Birth of the Clinic*, trans. Alan Sheridan, London:
Routledge, 1973. For a discussion, see Samuel Talcott, 'Georges
Canguilhem (1904–1995)', in Leonard Lawlor and John Nale
(eds.), *The Cambridge Foucault Lexicon*, Cambridge: Cambridge
University Press, 1994, 580–587.
48 Louis Althusser, 'A Letter to the Translator', *For Marx*,
trans. Ben Brewster, London: Verso, 1965, 257. See EGC 126–27
and Stefanos Geroulanos and Todd Meyers, 'Introduction:
Georges Canguilhem's Critique of Medical Reason', WM
–/6.
49 Louis Althusser, Étienne Balibar, Roger Establet, Pierre Macherey and Jacques Rancière, *Lire le Capital*, Paris: Presses Universitaires de France, 1996 [1965], vii, 7 n. 1; Althusser et al.,
Reading Capital: The Complete Edition, trans. Ben Brewster and
David Fernbach, London: Verso, 2016, 3, 14 n. 1. See also Althusser's introduction to Pierre Macherey, 'La philosophie de la
science de Georges Canguilhem', *La pensée* 113, 1964, 50–74;
translated in Macherey, *In a Materialist Way: Selected Essays by
Pierre Macherey*, ed. Warren Montag, trans. Ted Stolze, London:
Verso, 1998, 161–87. The essay, but not the introduction, is
reprinted in Pierre Macherey, *De Canguilhem à Foucault: La force
des normes,* Paris: La Fabrique, 2009.
50 Miller is Lacan's son-in-law, and editor of his seminars; Milner
is a linguist and author of *For the Love of Language*, trans. Ann
Banfield, Basingstoke: Macmillan, 1990.
51 For the *Cahiers,* see the reproduction of each issue at http://
cahiers.kingston.ac.uk, and for translations and supplementary
materials, see *Concept and Form Volume One: Key Texts from the
Cahiers pour l'Analyse* and *Concept and Form Volume Two: Interviews and Essays on the Cahiers pour l'Analyse*, ed. Peter Hallward
and Knox Peden, London: Verso, 2012.

52   https://dep-philo.parisnanterre.fr/les-enseignants/balibar-etienne-55434.kjsp; Lecourt, *Georges Canguilhem*, 117; Macherey, *De Canguilhem à Foucault*, 27; Samuel Lézé, 'Canguilhem', in Jean-Philippe Cazier (ed.), *Abécédaire de Pierre Bourdieu*, Sils Maria: Mons, 2007, 21–23.

53   Benoît Peeters, *Derrida: A Biography*, trans. Andrew Brown, Cambridge: Polity, 2013, 113–14, 129, 473.

54   Jacques Bouveresse, 'Préface aux *Oeuvres complètes* de Georges Canguilhem', in OC I, 7.

55   Reported in Marc Ragon, 'King Cang', *Libération* 4 February 1993, 19–21, 20.

56   David Macey, 'King Cang', *Radical Philosophy* 75, 1996, 52. See Mike Gane, *French Social Theory*, London: Sage, 2003, 169–70.

57   Gary Gutting, *Michel Foucault's Archaeology of Scientific Reason*, Cambridge University Press, 1989, ch. 1; Dominique Lecourt, *Marxism and Epistemology: Bachelard, Canguilhem and Foucault*, London: Verso, 1975. The latter includes translations of both *L'épistémologie historique de Gaston Bachelard*, Paris: Vrin, 1969, and *Pour une critique de l'épistémologie (Bachelard, Canguilhem, Foucault)*, Paris: François Maspero, 1972.

58   Alain Badiou, *Pocket Pantheon: Figures of Postwar Philosophy*, trans. David Macey, London: Verso, 2009, ch. 1; Pierre Bourdieu, *Sketch for a Self-Analysis*, trans. Richard Nice, Cambridge: Polity, 2007, esp. 26–30; Roberto Esposito, *Bíos: Biopolitics and Philosophy*, trans. Timothy Campbell, Minneapolis: University of Minnesota Press, 2008, 188–90; Gane, *French Social Theory*, 125–36; Hans-Jörg Rheinberger, *An Epistemology of the Concrete: Twentieth-Century Histories of Life*, Durham: Duke University Press, 2010, ch. 3; Élisabeth Roudinesco, *Philosophy in Turbulent Times: Canguilhem, Sartre, Foucault, Althusser, Deleuze, Derrida*, trans. William McCuaig, New York: Columbia University Press, 2008, ch. 1.

59   He is also discussed briefly in Nikolas Rose, *The Politics of Life Itself: Biomedicine, Power, and Subjectivity in the Twenty-First Century*, Princeton University Press, 2007, especially 41–4.

60   The most detailed study is Marina C. Brilman, 'Georges Canguilhem: Norms and Knowledge in the Life Sciences', unpublished Ph.D. thesis, London School of Economics, 2009.

61   François Dagognet, *Georges Canguilhem, philosophe de la vie*, Paris: Empêcheurs de penser rond, 1997; Macherey, *De Canguilhem à Foucault*; Claude Debru, *Georges Canguilhem, science et non-science*, Paris: Rue d'Ulm, 2004; Guillaume Le Blanc, *Canguilhem et les*

normes, Paris: Presses Universitaires de France, 1998; Le Blanc, *La vie humaine: Anthropologie et biologie chez Georges Canguilhem*, Paris: Presses Universitaires de France, 2002 (reissued as *Canguilhem et la vie humaine*, Paris: Presses Universitaires de France, 2010).

62 Guillaume Pénisson, *Le vivant et l'épistémologie des concepts: essai sur Le normal et le pathologique de Georges Canguilhem*, Paris: Harmattan, 2008; Élodie Giroux, *Après Canguilhem, définir la santé et la maladie*, Paris: Presses Universitaires de France, 2010; Cyriaque Geoffroy Ebissienine, *La problématique de la santé et de la maladie dans la pensée biomédicale: Essai sur la normalité biologique chez Georges Canguilhem*, Paris: Harmattan, 2010; Lucien R. Karhausen, *Dr Georges Canguilhem: Médicin anomal*, Paris: Harmattan, 2017.

63 Gilles Renard, *L'épistémologie chez Georges Canguilhem*, Paris: Nathan, 1996; Jacques Chatué, *Épistémologie et transculturalité Tome 2: Le paradigme de Canguilhem*, Paris: Harmattan, 2010.

64 As well as those already referenced, see François Bing, Jean-François Braunstein and Elisabeth Roudinesco (eds.), *Actualité de Georges Canguilhem: Le normal et le pathologique – actes du Xe Colloque de la Société internationale d'histoire de la psychiatrie et de la psychanalyse*, Paris: Synthélabo, 1998; Anne Fagot-Largeault, Claude Debru and Michel Morange (eds.), *Philosophie et médecine: En hommage à Georges Canguilhem*, Paris: Vrin, 2008. For a set of papers more loosely connected to his work, see Pierre F. Daled (ed.), *L'envers de la raison, alentour de Canguilhem*, Paris: Vrin, 2008.

65 *Revue de métaphysique et de morale* 90 (1), 1985, including 'Bibliographie des travaux de Georges Canguilhem', 99–105.

66 'Le normal et le pathologique en question', *Prospective et santé* 40, 1986/7; *Economy and Society* 27 (2–3), ed. Thomas Osbourne and Nikolas Rose, 1998, 151–248; 'Georges Canguilhem en son temps', *Revue d'histoire des sciences* 53 (1), 2000, 5–106; and 'Georges Canguilhem', *Dialogue: Canadian Philosophical Review* 52 (4), 2013, 617–723.

67 Lecourt, *Pour une critique de l'épistémologie*, 66; Lecourt, *Marxism and Epistemology*, 163.

68 See Stefanos Geroulanos and Daniela Ginsburg, 'Translators' Note', KL –/xv.

69 Something close to this can be found in Le Blanc, *La vie humaine*. See, especially, 16.

2   The Normal and the Pathological

1   Stefanos Geroulanos, *Transparency in Postwar France: A Critical History of the Present*, Stanford University Press, 2017, 199.

2   René Leriche, 'Équilibre de la santé et tempéraments', in *Encyclopédie française Tome VI: L'être humain*, Paris: Comité de l'Encyclopédie française, 1936, section 16, 1; quoted in NP 67/91.

3   André Lalande, *Vocabulaire technique et critique de la philosophie*, Paris: Presses Universitaires de France, 1972, 691, see 688.

4   Lalande, *Vocabulaire technique et critique de la philosophie*, 60–1.

5   Canguilhem references A. Juret, *Dictionnaire étymologique grec et latin*, Macon: Protat Frères, 1942.

6   Isidore Geoffroy Saint-Hilaire, *Histoire générale et particulière des anomalies de l'organisation chez l'homme et les animaux*, Paris: J.-B. Baillière, 1832, vol. I, 37; misquoted in NP 108–9/132–3.

7   Georges Teissier, 'En marge de l'Encyclopédie française: Une controverse sur l'évolution', *Revue trimestrielle de l'Encyclopédie française* 3, 1938, 11–14.

8   See his 1973 essay on this theme (IR 121–39/125–45).

9   Victor Prus, *De l'irritation et de la phlegmasie, ou nouvelle doctrine médicale*, Paris: Panckoucke, 1825, lii.

10   Claude Bernard, *Leçons sur la chaleur animale*, Paris: J.-B. Baillière, 1876, 57.

11   Pierre Vendryès, *Vie et probabilité*, Paris: Albin Michel, 1942.

12   Adolphe Quetelet, *Anthropométrie ou mesure des differentes facultés de l'homme*, Brussels: Muquardt, 1870, 22.

13   Maximilien Sorre, *Les fondements biologiques de la géographie humaine*, Paris: A. Colin, 1971 [1943].

14   Canguilhem returned to this theme in 'La question de l'écologie: La technique ou la vie', in François Dagognet, *Considérations sur l'idée de nature*, Paris: Vrin, 1990, 183–91.

15   The last clause is a quote from Léon Pales, *État actuel de la paléopathologie: Contribution à l'étude de la pathologie comparative*, Bordeaux: E. Drouillard, 1929, 307; Canguilhem also references Henry de Varigny, *La mort et la biologie*, Paris: Alcan, 1926.

16   Kurt Goldstein, *Der Aufbau des Organismus*, The Hague: Martinus Nijhoff, 1934; Goldstein, *The Organism: A Holistic Approach to Biology Derived from Pathological Data in Man*, New York: Zone, 1995 [1939]; Goldstein, 'L'analyse de l'aphasie et l'étude de l'essence du langage' (1933) in *Selected Papers / Ausgewählte Schriften*, ed. Aron Gurwitsch, E. M. Goldstein-Haudek and R. W. Haudek, The Hague: Martinus Nijhoff, 1971, 282–344.

17　Goldstein, *Der Aufbau des Organismus*, 265, 272; Goldstein, *The Organism*, 325, 330.

18　René Leriche, *La chirurgie de la douleur*, Paris: Masson, 1937; *The Surgery of Pain*, trans. Archibald Young, London: Baillière, Tindall and Cox, 1939.

19　Émile Durkheim, *Les règles de la méthode sociologique*, Paris: Felix Alcan, 1895; *The Rules of Sociological Method*, ed. George E. G. Catlin, trans. Sarah A. Solovay and John H. Mueller, New York: The Free Press, 1938, ch. 3. See Le Blanc, *La vie humaine*, 133–7.

20　Friedrich Nietzsche, *Kritische Studienausgabe*, ed. Giorgio Colli and Mazzino Montinari, 15 vols., Berlin: de Gruyter, 1999, vol. XIII, 250–1; Nietzsche, *The Will to Power*, ed. Walter Kaufmann, New York: Vintage, 1967, §47.

21　Bernard, *Leçons sur la chaleur animale*, 391; quoted in NP 47/71.

22　Esposito, *Bíos*, 188–9.

23　Geroulanos, *Transparency in Postwar France*, 216.

24　Claude Bernard, *Principes de médecine expérimentale*, Paris: Presses Universitaires de France, 1947. See also KL 155–69/121–33.

25　See Marina Brilman, 'Canguilhem's Critique of Kant: Bringing Rationality Back to Life', *Theory, Culture & Society* 35 (2), 2018, 25–46.

26　Hans Selye, *The Physiology and Pathology of Exposure to Stress: A Treatise Based on the Concepts of the General Adaptation Syndrome and the Diseases of Adaptation*, Montreal: Acta, 1959.

3　Philosophy of Biology

1　See KL 7/–; and Paola Marrati and Todd Meyers, 'Foreword: Life as Such', KL –/155 n. 1.

2　Esposito, *Bíos*, 188–9.

3　See Monica Greco, 'On the Vitality of Vitalism', *Theory, Culture & Society* 22 (1), 2005, 15–27.

4　Elsewhere, Canguilhem references Gilbert Simondon, *L'individu et sa genèse physico-biologique: L'individuation à la lumière des notions de forme et d'information*, Paris: Presses Universitaires de France, 1964 (KL 78 n. 1 / 166 n. 98).

5　As the translators note, 'the OED confirms the applicability in English of the double meaning, indeed places the interpretation of *engine* (or "engine") as "genius", "cunning", "trickery" or "evil machination" before its interpretation as machine' (KL – / 167 n. 3).

6   Emanuel Rädl, *Geschichte der biologischen Theorie in der Neuzeit*, Leipzig: Wilhelm Engelmann, 2nd edn, 1913, 147–9.

7   See note to the second edition (KL 93 n. 1* / 169 n. 18).

8   Bachelard, *Études*, 75–6, cited in KL 95/69.

9   Xavier Bichat, *Recherches physiologiques sur la vie et la mort*, Paris: Béchet, 3rd edn, 1805, 78 (article 7, §1), cited in KL 95/69.

10   Jean Giraudoux, *Three Plays, Vol. 2*, trans. Phyllis La Farge and Peter H. Judd, New York: Hill and Wang, 1964, 157–247.

11   Claude Bernard, *Introduction à la médecine expérimentale*, Paris: J.-B. Baillière et fils, 1865, 91; Bernard, *An Introduction to the Study of Experimental Medicine*, trans. Henry Copley Greene, New York: Henry Schuman, Inc, 1949, 93. The gloss on *methodus* comes from KL – / 159 n. 50.

12   Hans Driesch, *Die Überwindung des Materialismus*, Zürich: Rascher, 1935, 59. The sentence in question reads: 'Eine Maschine als *Werkzeug* für den Führer – aber *der Führer ist die Hauptsache*', 'A machine is a *tool* for a leader, but the leader is the main thing.' See also KL 63/42–3 on vitalism and totalitarianism.

13   Esposito, *Bíos*, 188–90.

14   On this, see KL 77/54, and Dominique Lecourt, *Proletarian Science? The Case of Lysenko*, trans. Ben Brewster, London: NLB, 1977.

15   Bernard, *Introduction à la médecine expérimentale*, 194; *An Introduction to the Study of Experimental Medicine*, 93.

16   It previously appeared in English as 'Machine and Organism', trans. Mark Cohen and Randall Cherry, in Jonathan Crary and Sanford Kwinter (eds.), *Incorporations*, New York: Zone Books, 1992, 44–69.

17   Giorgio Baglivi, *De praxi medica*, in *Opera omnia medico-practica et anatomica*, Venice, 1727, 78. He attributes a similar passage to François Magendie in EHPS 149; VR 267.

18   On Aristotle and Descartes, see also EHPS 323–5; VR 294–6.

19   Pierre-Maxime Schuhl, *Machinisme et philosophie*, Paris: Alcan, 1938; Lucien Laberthonnière, *Oeuvres: Études sur Descartes*, ed. Louis Canet, 2 vols., Paris: Vrin, 1935; Franz Borkenau, *Der Übergang vom feudalen zum bürgerlichen Weltbind: Studien zur Geschichte der Philosophie der Manufakturperiode*, Paris: Alcan, 1971 [1934]. See also OC I, 490–6; VR 219–26.

20   Henryk Grossmann, 'Die gesellschaftlichen Grundlagen der mechanistischen Philosophie und die Manufaktur', *Zeitschrift für Sozialforschung* 4, 1935, 161–231. On da Vinci, see also the discussion in relation to Vesalius in EHPS 32–3 and FCR 21.

21 Geroulanos and Ginsburg, 'Translators' Note', KL –/xiv.
22 These can be found in *The Philosophical Writings of Descartes*, trans. John Cottingham, Robert Stoothoff, Dugald Murdoch and Anthony Kenny, 3 vols., Cambridge University Press, 1985–91; Descartes, *Treatise of Man*, French–English edition, trans. Thomas Steele Hall, Cambridge, MA: Harvard University Press, 1972.
23 Nicolaus Steno, 'The Discourse on the Anatomy of the Brain', in Troels Kardel and Paul Maquet (eds.), *Nicolaus Steno: Biography and Original Papers of a 17th Century Scientist*, Berlin: Springer, 2013, 507–27, praises Descartes's work on anatomy, but recognizes that his account of the body is not the same as that of the anatomist (KL 113/85–6). An excerpt is provided as Appendix 3 of KL 188–9/152–3.
24 Claude Bernard, *Leçons sur les phénomènes de la vie communs aux animaux et aux végétaux*, 2 vols., Paris: J.-B. Baillière, 1878, vol. I, 51; Bernard, *Lectures on the Phenomena of Life Common to Animals and Plants*, trans. Hebbel E. Hoff, Roger Guillemin, and Lucienne Guillemin, Springfield, IL: Charles C. Thomas, 1974, 36; see KL 114–15/87.
25 Alfred Espinas, *Étude sociologique: Les origines de la technologie*, Paris: Félix Alcan, 1897; Ernst Kapp, *Grundlinien einer Philosophie der Technik*, Braunschweig: George Westermann, 1877.
26 André Leroi-Gourhan, *Milieu et techniques: Évolution et techniques II*, Paris: Albin Michel, 1992 [1945].
27 Georges Friedmann, *Machine et humanisme II: Problèmes humains du machinisme industriel*, Paris: Gallimard, 1946; Friedmann, *Industrial Society: The Emergence of the Human Problems of Automation*, ed. Harold L. Sheppard, Glencoe, IL: Free Press, 1955. See Canguilhem's review 'Milieu et norms de l'homme au travail', *Cahiers internationaux de sociologie* 23, 1947, 120–36, reprinted in OC IV, 291–306.
28 As with so many of Canguilhem's other texts, it derived from his Paris teaching. Notes for 'Modèle et analogie en physique' are filed as 'L'analogue et le singulier dans la science', CAPHÉS GC 13.3.1, 75–81.
29 See G. W. F. Hegel, *Science of Logic*, trans. A.V. Miller, Amherst: Humanity Books, 1969, 'The Relation of Whole and Parts', 513–17.
30 See Geroulanos and Ginsburg, 'Translators' Note', KL –/xiv.
31 Hippolyte Taine, *Essais de critique et d'histoire*, Paris: Hachette, 1904.

32   Montesquieu, *The Spirit of the Laws*, ed. and trans. Anne M. Cohler, Basia Carolyn Miller and Harold Samuel Stone, Cambridge University Press, 1989, Part II, Books 14–18.

33   Auguste Comte, *Philosophie première: Cours de philosophie positive, Leçons 1 à 45*, Paris: Hermann, 1975, 682 n. *; cited in KL 133/101.

34   Paul Tannery, 'Auguste Comte et l'histoire des sciences', *Revue Générale des Sciences Pures et Appliqués* 16, 1905, 410–17.

35   'Histoire de l'homme et nature des choses selon Auguste Comte', *Les Etudes philosophiques* 3, 1974, 293–7, 294.

36   'Histoire de l'homme et nature des choses', 295.

37   Carl Ritter, *Comparative Geography*, trans. William L. Gage, Edinburgh: William Blackwood and Sons, 1865; Alexander von Humboldt, *Kosmos: Entwurf einer physischen Weltbeschreibung*, 5 vols., Stuttgart: Cotta, 1845–62.

38   On the history of geography, a useful initial source is David Livingstone, *The Geographical Tradition: Episodes in the History of a Contested Enterprise*, Oxford: Blackwell, 1992.

39   See, in particular, Lucien Febvre, *La terre et l'évolution humaine: Introduction géographique à l'histoire*, Paris: La Renaissance du livre, 1922; translated by E. G. Mountford and J. H. Paxton as *A Geographical Introduction to History*, London: Routledge and Kegan Paul, 1925.

40   Neither Canguilhem nor his translators provide a reference to Koffka's work to substantiate this claim. Instead, Canguilhem references Paul Guillaume, *La psychologie de la forme*, Paris: Flammarion, 1937; and Maurice Merleau-Ponty, *La structure du comportement*, Paris: Presses Universitaires de France, 1942; translated by Alden L. Fisher as *The Structure of Behavior*, Boston: Beacon Press, 1963.

41   See Goldstein, *Der Aufbau des Organismus*, 58–64; *The Organism*, 85–90. On this relation, see Debru, *Georges Canguilhem*, ch. III.

42   Jakob von Uexküll, *A Foray into the Worlds of Animals and Humans*, trans. Joseph O'Neill, Minneapolis: University of Minnesota Press, 2010.

43   See Louis Bounoure, *L'autonomie de l'être vivant: Essais sur les formes organiques et psychologiques de l'activité vitale*, Paris: Presses Universitaires de France, 1949, 143–4.

44   See François Dagognet, *Philosophie biologique*, Paris: Presses Universitaires de France, 1955, 94–7.

45   Albert Brachet, *La vie créatrice des formes*, Paris: Alcan, 1927, 171.

46   Alexandre Koyré, *The Astronomical Revolution: Copernicus, Kepler, Borelli*, trans. R. E. W. Maddison, London: Methuen, 1973; and

Koyré, *From the Closed World to the Infinite Universe*, Baltimore: Johns Hopkins University Press, 1957.

4 Physiology and the Reflex

1 Marrati and Meyers, 'Foreword', in KL –/xi.
2 Leroi-Gourhan, *Milieu et techniques*, 370, quoted in KL 10/xviii.
3 Michel Foucault, *Les mots et les choses: Une archéologie des sciences humaines*, Paris: Gallimard/Tel, 1966; Foucault, *The Order of Things: An Archaeology of the Human Sciences*, trans. Alan Sheridan, London: Routledge, 1970.
4 R. A. Crowson, 'Darwin and Classification', in S. A. Barnett (ed.), *A Century of Darwin*, London: Heinemann, 1958, 102–29, 122.
5 Goldstein, *Der Aufbau des Organismus*, 340; Goldstein, *The Organism*, 378.
6 Geroulanos and Ginsburg, 'Translators' Note', KL –/xiii–xiv.
7 On Bernard, see also 'Théorie et technique de l'expérimentation chez Claude Bernard', in Étienne Wolff, Christian Fouchet, Bernard A. Houssay et al. *Philosophie et méthodologie scientifiques de Claude Bernard*, Paris: Masson & Cie, 1967, 23–32; and his preface to Bernard, *Leçons sur les phénomènes*, 7–14.
8 Claude Bernard, *Leçons de physiologie experimentale appliquée à la médecine*, 2 vols., Paris: J-B. Baillière et fils, 1855–6.
9 Bernard, *Introduction à la médecine expérimentale*, 90; Bernard, *Introduction to the Study of Experimental Medicine*, 51.
10 'Un physiologiste philosophe: Claude Bernard', *Dialogue: Canadian Philosophical Review* 5 (4) 1967, 555–72, 566. On Bichat, see EHPS 156–62.
11 'Un physiologiste philosophe', 567.
12 Mirko Drazen Grmek, 'Réflexions inédites de Claude Bernard sur la médecine pratique', *Médecine en France* 150, 1964, 6–11, 7; cited in EHPS 394; VR 283.
13 See also 'Préface', in Bernard, *Leçons sur les phénomènes*, vol. I, 9.
14 Bernard, *Leçons de physiologie expérimentale*, vol I, 54; quoted in EHPS 146; VR 264.
15 'Physical vitalism' appears on the final page of the text: Bernard, *Leçons sur les phénomènes de la vie*, vol. II, 524.
16 On ablation of glands, see 'Physiologie en Allemagne', OC IV, 996–8; VR 106.

17 The initial manuscript and typescript can be found in CAPHÉS GC 7.1.
18 See Debru, *Georges Canguilhem, science et non-science*, 18.
19 Jean-François Braunstein, 'Psychologie et milieu: Éthique et histoire des sciences chez Georges Canguilhem', in Braunstein (ed.), *Canguilhem*, 63–89, 76.
20 Braunstein, 'Psychologie et milieu', 80.
21 Franklin Fearing, *Reflex Action: A Study in the History of Physiological Psychology*, Cambridge, MA: MIT Press, 1970 [1930].
22 Charles Scott Sherrington, *The Integrative Action of the Nervous System*, New Haven, CT: Yale University Press, 1947 [1906]. Canguilhem does recognize that this is not the case in Sherrington, *The Endeavour of Jean Fernel, with a List of the Editions of his Writings*, Cambridge University Press, 1946.
23 Hebbel E. Hoff and Peter Kellaway, 'The Early History of the Reflex', *Journal of the History of Medicine and Allied Sciences* 7 (2), 1952, 211–49.
24 J. F. Fulton, 'Historical Introduction', in *Muscular Contraction and the Reflex Control of Movement*, London: Baillière, Tindall and Cox, 1926, 3–55; *Physiology of the Nervous System*, Oxford University Press, 3rd edn, 1949 [1938] (historical notes precede each chapter).
25 Fearing, *Reflex Action*, 26 and n. 12.
26 Sherrington, *The Endeavour of Jean Fernel*, 84, cited in FCR 37 and n. 3.
27 Thomas Willis, 'De motu musculari: Exercitatio medico-physica', in *Opera omnia*, Geneva: Samuelem de Tournes, 1695, vol. I, 42–3.
28 For a succinct summary of these claims, see OC IV, 805–12; partly translated in VR 189–93.
29 Gaston Bachelard, *L'activité rationaliste de la physique contemporaine*, Paris: Presses universitaires de France, 25–6.
30 Viktor von Weizsäcker, 'Geleitwort zu einer Abhandlung von Ernst Marx', in *Gesammelte Schriften*, Frankfurt am Main: Suhrkamp, 1997, vol. IV, 72.
31 Gaston Bachelard, *La psychanalyse du feu*, Paris: Gallimard, 1949 [1938]; Bachelard, *The Psychoanalysis of Fire*, Boston: Beacon, 1977; Bachelard, *La dialectique de la durée*, Paris: Presses Universitaires de France, 1963 [1936]; Bachelard, *La terre et les rêveries du repos: Essai sur les images de l'intimité*, Paris: José Corti, 1948; translated by Mary McAllestor Jones, *Earth and Reveries of Repose:*

An Essay on Images of Interiority, Dallas Institute Publications, 2011.

32 It only appears in the bibliography of E. G. T. Liddell, *The Discovery of Reflexes*, Oxford: Clarendon Press, 1960, 145.

33 See, for example, Richard Lowry, 'The Reflex Model in Psychology: Origins and Evolution', *Journal of the History of Behavioral Sciences* 6, 1970, 64–9; Thomas S. Hall, 'Descartes' Physiological Method', *Journal of the History of Biology* 3, 1970, 53–79.

34 The essay is reprinted in EHPS 295–304; part-translated in VR 195–202. See also EHPS 310; RAM 512, which makes brief mention of reflex arcs and reflex actions, with reference to FCR.

35 Reinhard Koselleck, *Futures Past: On the Semantics of Historical Time*, trans. Keith Tribe, New York: Columbia University Press, 2004, 88; and see Koselleck, *The Practice of Conceptual History: Timing History, Spacing Concepts*, Stanford University Press, 2002. On Canguilhem's approach, see Henning Schmidgen, 'The Life of Concepts: Georges Canguilhem and the History of Science', *History and Philosophy of the Life Sciences* 36 (2), 2014, 232–53.

36 Jean Fernel, *The Physiologia of Jean Fernel (1567)*, Latin–English edn, trans. J. M. Forrester, Philadelphia: American Philosophical Society, 2003, 601. Canguilhem renders this as the 'nature of the healthy man, of all his forces and all his functions'.

37 Claude Bernard, *Rapport sur les progrès et la marche de la physiologie générale en France*, Paris: L'imprimerie Impériale, 1867, 131.

5 Regulation and Psychology

1 The lecture is included in *Writings on Medicine*, without the discussion. But it is a slightly odd choice to include since it dates from a generation before the other pieces in that collection, and Canguilhem revisited its themes in later work.

2 Geroulanos and Meyers, 'Introduction', WM –/20.

3 Pierre-Maxime Schuhl, *Essai sur la formation de la pensée grecque: Introduction historique à une étude de la philosophie platonicienne*, Paris: F. Alcan, 1934, 309.

4 Ernst Haeckel, 'Cell-Soul, Soul-Cells', in *The Pedigree of Man and Other Essays*, trans. Edward B. Aveling, London: Freethought, 1883, 135–74, 172: 'The government of the animal body is a cell-monarchy, that of the plant a cell-republic. As all the

separate cells in a plant remain much more independent than in an animal, the unity of the soul is much less possible in the former than it is in the latter.' See Ernst Haeckel, *Die Welträtsel: Gemeinverständliche Studien über monistische Philosophe*, Berliner Ausgabe, 2016 [1899], 27, discussed in KL 70/48.

5   Geroulanos and Meyers, WM –/99 n. 6.

6   Canguilhem's reference is Walter B. Cannon, *The Wisdom of the Body*, New York: Norton, 2nd edn, 1963 [1932], but, as his translators note, the term first appears in 'Physiological Regulation of Normal States: Some Tentative Postulates Concerning Biological Homeostatics' (1926), reprinted in L. L. Langley (ed.), *Homeostasis: Origins of the Concept*, Stroudsburg: Dowden, Hutchinson & Ross, 1973, 246–9. Bernard, *Introduction à l'étude de la médecine expérimentale*, 112; Bernard, *An Introduction to the Study of Experimental Medicine*, 119.

7   Bernard, *Introduction à l'étude de la médecine expérimentale*, 66; Bernard, *An Introduction to the Study of Experimental Medicine*, 64.

8   'Un physiologiste philosophe', 567. On internal milieu as blood, see EHPS 245; on endocrinology – internal secretions by gland to regulate the body – see EHPS 262–5; VR 118–22; EHPS 128–9, and EHPS 147–8; VR 268.

9   Henri Bergson, *The Two Sources of Morality and Religion*, trans. Cloudesley Brereton, R. Ashley Audra and W. Horsfall Carter, London: Macmillan, 1935.

10 'Régulation (épistemologie)', *Encyclopædia Universalis* 14, 1972, 1–3.

11 Antoine Lavoisier, *Oeuvres*, ed. J.-B. Dumas, E. Grimaux and F.-A. Fouqué, Paris, 1862, vol. II, 700.

12 'Régulation (épistemologie)', 2.

13 André Lichnerowicz, François Perroux and Gilbert Gadoffre (eds.), *L'idée de régulation dans les sciences: Seminaires interdisciplinaires du Collège de France*, Paris: Maloine, 1977.

14 Geroulanos and Meyers, 'Introduction', WM –/21. 'Fin des normes ou crise des regulations? 3 conférences à Louvain, 20–22 mars 1973', CAPHÉS GC 25.17, subfolder 1.

15 'Cours, séminaire, notes et coupures de presse sur le "sauvage". 1959–1974', CAPHÉS GC 25.17, subfolder 2.

16 'La régulation comme réalité et comme fiction', CAPHÉS GC 25.17, subfolder 1, 93.

17 Hans Driesch, *Die organischen Regulationem: Vorbereitungen zu einer Theorie des Lebens*, Leipzig: Wilhelm Engelmann, 1901.

18   Alexandre Koyré, *Études newtoniennes*, Paris: Gallimard, 1968, 40.

19   Michel Foucault, *L'archéologie du savoir*, Paris: Gallimard/Tel, 2014 [1969], 197; Foucault, *The Archaeology of Knowledge*, trans. A. M. Sheridan-Smith, London: Routledge, 1972, 162.

20   The passage is from Gottfried Leibniz, 'Responsiones ad Stahlianas observationes', in *Opera omnia*, ed. Ludovici Dutens, 7 vols., Geneva: Fratres de Tournes, 1768, vol. II, part II, 149.

21   François Jacob, *La logique du vivant: Une histoire de l'hérédité*, Paris: Gallimard, 1970; Jacob, *The Logic of Life: A History of Heredity and The Possible and the Actual*, trans. Betty E. Spillmann, London: Penguin, 1989. See Samuel Talcott, 'Errant Life, Molecular Biology, and Biopower: Canguilhem, Jacob and Foucault', *History and Philosophy of the Life Sciences* 36 (2), 2014, 254–79.

22   See Comte, *Philosophie première*, 680–1.

23   Auguste Comte, *Système de politique positive ou Traité de sociologie*, Paris: Carilian-Gœury, 1852, vol. II, 26; cited in IR 95/96.

24   Bernard, *Rapport sur les progrès et la marche*, 182, cited in IR 96/97.

25   Mirko Drazen Grmek, 'Évolution des conceptions de Claude Bernard sur le milieu intérieur', in Wolff et al., *Philosophie et méthodologie scientifiques de Claude Bernard*, 117–50, 140.

26   Braunstein, 'Psychologie et milieu', 71.

27   Braunstein, 'Psychologie et milieu', 65.

28   Braunstein, 'Psychologie et milieu', 89.

29   Pinel is discussed in more detail in Foucault, *History of Madness*. See also Othmar Keel, *La généalogie de l'histopathologie: Une revision déchirante – Philippe Pinel, lecteur discret de J.-C. Smyth (1741–1821)*, Paris: Vrin, 1979. Canguilhem contributes a brief preface on I–II.

30   Stendhal, *Mémoires d'un touriste*, Paris: Michel Lévy frères, 1854, vol. II, 23.

31   Alfred C. Kinsey, Wardell B. Pomeroy and Clyde E. Martin, *Sexual Behavior in the Human Male*, Philadelphia: W. B. Saunders, 1948; and Kinsey et al., *Sexual Behavior in the Human Female*, Philadelphia: W. B. Saunders, 1953.

32   See, for example, Jacques Lacan, *Écrits,* Paris: Seuil, 1999 [1966], vol. II, 339–40, where it becomes a toboggan going downhill; Rose, *The Politics of Life Itself*, 195.

33   Canguilhem told this to David Macey, reported in *The Lives of Michel Foucault*, London: Hutchinson, 1993, 104.

34 'Qu'est-ce que la psychologie?' *Cahiers pour l'analyse* 2 (1), 1966, 77–91. It was followed by Robert Pagès, 'Quelques remarques sur "Qu'est-ce que la psychologie?"', 92–7, and Canguilhem, 'Note', 98.

35 F. J. Gall and G. Spurzheim, *Recherches sur le système nerveux en général, et sur celui du cerveau en particulier*, Paris: F. Schoell and H. Nicolle, 1809.

36 Georges Lantéri-Laura, *Histoire de la phrénologie: L'homme et son cerveau selon F. J. Gall*, Paris: Presses Universitaires de France, 1970.

37 See Sigmund Freud, 'The Unconscious', in *On Metapsychology: The Theory of Psychoanalysis*, London: Penguin, 1991, 167–222.

38 Pierre Janet, reported by Marcel Jousse, *Études de psychologie linguistique: Le style oral rythmique et mnémotechnique chez les verbo-moteurs*, Paris: Gabriel Beauchesne, 1925, 39.

39 Pierre Janet, *La pensée intérieure et ses troubles: Leçons au Collège de France 1926–1927*, Paris: L'Harmattan, 2007, 361; see BT 16/9.

40 Henri Altan, *Entre le crystal et la fumée: Essais sur l'organisation du vivant*, Paris: Seuil, 1979, 221; cited in BT 22/11.

41 Massimo Piattelli-Palmarini (ed.), *Language and Learning: The Debate between Jean Piaget and Noam Chomsky*, Cambridge, MA: Harvard University Press, 1980.

42 Edmund Husserl, *Philosophie première 1: Histoire critique des idées*, Paris: Presses Universitaires de France, 1970 [1923–4], 75.

6   Evolution and Monstrosity

1 Among other works, see Jacques Ulmann, *De la gymnastique aux sports modernes: Histoire des doctrines de l'éducation physique*, Paris: Presses Universitaires de France, 1965; Ulmann, *La nature et l'éducation: L'idée de nature dans l'education, physique et morale*, Paris: Klincksieck, 1987; Ulmann, *Corps et civilisation: Éducation physique, médecine, sport*, Paris: Vrin, 1993.

2 Georges Lapassade, *L'entrée dans la vie*, Paris: Minuit, 1963; Lapassade, *Groupes, organisations, institutions*, Paris: Gauthiers-Villars, 1975; Lapassade, *Les états modifiés de la conscience*, Paris: Presses Universitaires de France, 1987.

3 Jacques Piquemal, 'Le choléra de 1832 en France et la pensée médicale', *Thalès* 10, 1959, 27–73; Piquemal, *Aspects de la pensée de Mendel*, Paris: Palais de la découverte, 1965; Piquemal, *Essais*

et leçons d'histoire de la médecine et de la biologie, Paris: Presses Universitaires de France, 1993.

4   Caspar Friedrich Wolff, *Theoria generationis*, Halle: Hendel, 1774; Charles Darwin, *On the Origin of Species by Means of Natural Selection*, London: John Murray, 1859.

5   They reference a Russian study from 1961, but say that they regret not taking it into account (DE 121).

6   He is briefly mentioned in relation to Darwin in DE 70, 75, and more extensively in EHPS 211–25. Canguilhem elsewhere approvingly cites Henri Daudin, *Cuvier et Lamarck: Les classes zoologiques et l'idée de série animale (1790–1830)*, 2 vols., Paris: Vrin, 1926 (EHPS 101).

7   James Barr, 'Why the World was Created in 4004 BC', *Bulletin of the John Rylands University Library* 67, 1985, 575–608.

8   Crowson, 'Darwin and Classification', 122: 'In his personal character and relations, Darwin stands out in later nineteenth-century England, not as the vanguard of twentieth-century biology, but as one of his own "living fossils", an eighteenth-century savant living in the railway age.'

9   Charles Darwin, *The Descent of Man, and Selection in Relation to Sex*, 2 vols., London: John Murray, 1871.

10 Thomas Henry Huxley, 'Evolution in Biology', in *Darwiniana: Collected Essays, Volume II*, London: Macmillian & Co., 1893, 187–226.

11 Étienne Balibar and Dominique Lecourt, 'Présentation', in DE 5.

12 Balibar and Lecourt, 'Présentation', in DE 6.

13 Balibar and Lecourt, 'Présentation', in DE 6–7.

14 'Georges Cuvier', *Revue d'histoire des sciences et de leurs applications* 23 (1), 1970.

15 Canguilhem is thinking of Limoges's doctorate on 'La constitution du concept darwinien de sélection naturelle' (EHPS 110 n. 21), which was published in Canguilhem's 'Galien' series as *La sélection naturelle: Étude sur la première constitution d'un concept*, Paris: Presses Universitaires de France, 1970. Gillespie was an American historian of science, best known for *Genesis and Geology: A Study in the Relations of Scientific Thought, Natural Theology, and Social Opinion in Great Britain, 1790–1850*, Cambridge, MA: Harvard University Press, 1996 [1951].

16 'Histoire de la tératologie depuis Étienne Geoffroy Saint-Hilaire, 1961–62', CAPHÉS GC 12.2.8.

17   KL 236 n. * / 182 n. 31. There is some discussion of monstrosity in the co-authored study DE 25–6, 29–30, also published in 1962.

18   The reference appears to be a modification of Bachelard, *Études*, 85: 'I am the limit of my lost illusions.' *Illusions perdues* is the title of a novel by Balzac. See also OC IV, 734.

19   Gabriel Tarde, *L'opposition universelle: Essai d'une théorie des contraires*, Paris: F. Alcan, 1897, 25.

20   Ernest Martin, *Histoire des monstres depuis l'Antiquité jusqu'à nos jours*, Grenoble: Millon, 2002 [1880].

21   Foucault would return to these themes in more detail in some of his lecture courses, notably *Abnormal: Lectures at the Collège de France 1974–75*, trans. Graham Burchell, London: Verso, 2003. See Stuart Elden, *Foucault's Last Decade*, Cambridge: Polity, 2016, ch. 1; and François Delaporte, 'Foucault, Canguilhem et les monstres', in Braunstein (ed.), *Canguilhem*, 91–112.

22   Gottfried Leibniz, *New Essays on Human Understanding*, ed. and trans. Peter Remnant and Jonathan Bennett, Cambridge University Press, 1996, book III, ch. 6, §§ 12, 14, 17, and book IV, ch. 4, §13.

23   On the link between teratology and embryology, see also 'Histoire de l'homme et nature des choses', 297; VR 251.

24   Camille Dareste, *Recherches sur la production artificielle des monstruosités; ou, Essais de tératogénie expérimentale*, Paris: Reinwald, 2nd edn, 1891 [1877], 25.

25   'L'analogue et le singulier dans la science', CAPHÉS GC 13.3.1. A summary of 'Du singulier et de la singularité en épistémologie biologique, Société belge de philosophie, 10 février 1962' is found in the same folder, 116–17.

26   Aristotle, *Politics*, Greek–English edition, trans. H. Rackham, London: William Heinemann, 1932, 1253a9; see Aristotle, *Nicomachean Ethics*, Greek–English edition, trans. H. Rackham, London: William Heinemann, 1926, 1098a3–5, 1139a5–6.

7   Philosophy of History

1   Badiou, *Pocket Pantheon*, ch. 1, looks at Canguilhem and Cavaillès through the lens of Canguilhem's book.

2   For the details, the standard source is the biography by his sister: Gabrielle Ferrières, *Jean Cavaillès, philosophe et combatant, 1903–1944*, Paris: Presses Universitaires de France, 1950;

re-edited as *Jean Cavaillès, un philosophe dans la guerre, 1903–1944*, Paris: Seuil, 1982; translated by T. N. F. Murtagh as *Jean Cavaillès: A Philosopher in Time of War, 1903–1944*, Lampeter: Edwin Mellon Press, 2000. References are to the second French edition and the English, separated by '/'. The French also includes an essay by Gaston Bachelard, 'L'oeuvre de Jean Cavaillès', 209–21, that is not in the English translation.

3   https://www.unistra.fr/index.php?id=jean-cavailles. There is also a room named after him at the ENS.

4   Ferrières, *Jean Cavaillès*, 76/66.

5   Edmund Husserl, *Cartesian Meditations: An Introduction to Phenomenology*, trans. Dorian Cairns, The Hague: Martinus Nijhoff, 1960.

6   Jean Cavaillès, *Oeuvres complètes de philosophie des sciences*, Paris: Hermann, 1994.

7   See, in particular, Jean Cavaillès, *Méthode axiomatique et formalisme: Essai sur le problème du fondement des mathématiques*, Paris: Hermann, 1938; and Cavaillès, *Remarques sur la formation de la théorie abstraite des ensembles: Étude historique et critique*, Paris: Hermann, 1938 – both reprinted in his *Oeuvres complètes*.

8   'Une vie, une oeuvre 1903–1944', in Cavaillès, *Oeuvres complètes de philosophie des sciences*, 683–6.

9   Jean Cavaillès, *Sur la logique et la théorie de la science*, ed. Georges Canguilhem and Charles Ehresmann, Paris: Presses Universitaires de France, 1960 [1947]. There is a brief preface by Gaston Bachelard (v–vii) and an 'avertissement' by Canguilhem and Ehresmann (ix–xii, reprinted in OC IV, 244–7). Canguilhem edited another short text by Cavaillès, 'Mathématiques et formalisme', *Revue internationale de philosophie* 3 (8), 1949, 158–65, with a short introduction (OC IV, 359–60).

10 Cavaillès, *Sur la logique et la théorie de la science*, 78.

11 Ferrières, *Jean Cavaillès*, 183/175.

12 Limoges, OC IV, 223 n. 1, 260 n. 1. It was previously thought he was killed in January.

13 Gaston Bachelard, *Le matérialisme rationnel*, Paris: Presses Universitaires de France, 1953, 207–24.

14 'Préface', in Bachelard, *L'engagement rationaliste*, 6. See *Études*; *Le droit du rêver*, Paris: Presses Universitaires de France, 1970.

15 For good discussions, see, among others, Maurice Lalonde, *La théorie de la connaissance scientifique selon Gaston Bachelard*, Ottawa: Fides, 1964; Lecourt, *L'épistémologie historique de Gaston Bachelard*, translated in *Marxism and Epistemology*; Lecourt,

Bachelard ou le jour et la nuit, Paris: Bernard Grasset, 1974; Cristina Chimisso, *Gaston Bachelard: Critic of Science and the Imagination*, Abingdon: Routledge, 2001; Zbigniew Kotowicz, *Gaston Bachelard: A Philosophy of the Surreal*, Edinburgh University Press, 2016.

16 Gaston Bachelard, *La formation de l'esprit scientifique: Contribution à une psychanalyse de la connaissance objective*, Paris: Vrin, 5th edn, 1967 [1934]; Bachelard, *The Formation of the Scientific Mind*, trans. Mary McAllester Jones, Bolton: Clinamen, 2002; Bachelard, *Dialectic of Duration*, trans. M. McAllester Jones, Bolton: Clinamen, 2000; Bachelard, *L'intuition de l'instant*, Paris: Éditions Gonthier, 1932; Bachelard, *Intuition of the Instant*, trans. Eileen Rizo-Patron, Evanston, IL: Northwestern University Press, 2013.

17 Gaston Bachelard, *Essai sur la connaissance approchée*, Paris: Vrin, 3rd edn, 1969 [1927]; and Bachelard, *Étude sur l'évolution d'un problème de physique: la propagation thermique dans les solides*, Paris: Vrin, 2nd edn, 1973 [1927].

18 Foucault, *Dits et écrits*, vol. IV, 768–9; Foucault, *Essential Works*, vol. II, 470. This is a revised version of an introduction to NP –/7–24.

19 See 'Preface', *History and Technology* 4 (1), 1987, 7–10.

20 Alexandre Koyré, *Études galiléennes*, Paris: Hermann, 1939.

21 See also Michel Foucault, *L'ordre du discours*, Paris: Gallimard, 1970, 36; translated by Ian McLeod as 'The Order of Discourse', in Robert Young (ed.), *Untying the Text: A Post-Structuralist Reader*, London: Routledge, 1981, 48–78, 60.

22 Alexandre Koyré, in 'Séance du 25 avril 1936: Sur la notion d'histoire de la philosophie', *Bulletin de la Société française de philosophie* 36 (3), 1936, 103–40, 136; quoted in 'Histoire de la philosophie et histoire des sciences, mars 1945', CAPHÉS GC 11.3.10, 6. See Geroulanos and Meyers, 'Introduction', WM – / 80 n. 13.

23 Gutting, *Michel Foucault's Archaeology of Scientific Reason*, 32.

24 Lecourt, *Pour une critique de l'épistémologie*, 69; *Marxism and Epistemology*, 166.

25 On the relation, see Alison Ross and Amir Ahmadi, 'Gaston Bachelard (1884–1962) and Georges Canguilhem (1904–1995): Epistemology in France', in Julian Wolfreys (ed.), *Modern European Criticism and Theory : A Critical Guide*, Edinburgh University Press, 2006, 90–7; and Cristina Chimisso, 'The Life Sciences and

French Philosophy of Science: Georges Canguilhem on Norms',
in Hanne Andersen, Dennis Dieks, Wenceslao J. Gonzalez,
Thomas Uebel and Gregory Wheeler (eds.), *New Challenges to
Philosophy of Science*, Berlin: Springer, 2013, 399–409.

26 Geroulanos and Ginsburg, 'Translators' Note', KL –/xiii; OC
IV, 249.

27 Bernard, *Introduction à l'étude de la médecine expérimentale*, 132;
Bernard, *An Introduction to the Study of Experimental Medicine*,
142.

28 No reference is provided, but this is a major theme of Bachelard,
*La formation de l'esprit scientifique / The Formation of the Scientific
Mind*. See, for example, 16/28.

29 'Tissue is made of thread [*fil*], that is to say, originally, of plant
fibres. That this word *fil* connotes images of continuity comes
across in expressions such as *fil d'eau* [water current] and *fil du
discours* [thread of an argument]' (KL 64 n. 1 / 163 n. 64). On
resistance to microscopes, see also KL 52/33 (Buffon) and IR
114/117 (Bernard).

30 See Gaston Bachelard, *Le nouvel esprit scientifique*, Paris: Presses
Universitaires de France, 1968 [1934]; Bachelard, *The New Scientific Spirit*, trans. Arthur Goldhammer, Boston: Beacon Press,
1984.

31 See A. R. Hall, *Philosophers at War: The Quarrel between Newton
and Gottfried Leibniz*, Cambridge University Press, 1980.

32 Eduard Jan Dijksterhuis, 'The Origins of Classical Mechanics
from Aristotle to Newton', in Marshall Clagett (ed.), *Critical
Problems in the History of Science*, Madison: University of Wis-
consin Press, 1962, 163–84, 182; quoted in EHPS 12; VR 42.
Canguilhem returns to this in IR 12–13/2–3.

33 Dijksterhuis, 'The Origins of Classical Mechanics', 182.

34 Pierre Lafitte, *Cours sur l'histoire générale des sciences: Discours
d'ouverture*, Paris: Revue occidentale, 1892, 24; quoted in EHPS
12; VR 43.

35 Althusser, 'A Letter to the Translator', 257.

36 Joseph T. Clark, SJ, 'The Philosophy of Science and the History
of Science', in Claget (ed.), *Critical Problems in the History of
Science*, 103–40, 103, speaks of a '*precursitis* virus'.

37 Foucault, *Les mots et les choses*, 158–76; *The Order of Things*,
145–62, especially 166/153.

38 Foucault, *L'archéologie du savoir*, 252–6; *The Archaeology of Knowledge*, 205–8.

39 Foucault, *L'archéologie du savoir*, 254; *The Archaeology of Knowledge*, 207.

40 See, for example, Foucault, *L'ordre du discours*, 73–4; 'The Order of Discourse', 73–4; Foucault, *L'archéologie du savoir*, 11, 195 n. 1, 236, 257–9; Foucault, *The Archaeology of Knowledge*, 5, 160 n. 1, 191–2, 209. As the last of these passages indicates, Foucault's project operates at a deeper level than Canguilhem – archaeology, not the history, of science. On these tensions, more generally, see Alison Ross, 'The Errors of History: Knowledge and Epistemology in Bachelard, Canguilhem and Foucault', *Angelaki* 23 (2), 2018, 139–54, who sees their approaches to knowledge and epistemology as 'fundamentally irreconcilable' (144). There is unfortunately little about Canguilhem in David Webb's excellent *Foucault's Archaeology: Science and Transformation*, Edinburgh University Press, 2013. He says this is because the connection 'is already well documented', though that might be a more open question.

41 Suzanne Bachelard, 'Épistémologie et histoire des sciences', in *Revue de Synthèse: XIIe Congrès international d'histoire des sciences, Colloques, textes des rapports*, Paris: Albin Michel, 1968, 39–51.

42 Suzanne Bachelard, *La conscience de la rationalité: Étude phénoménologique sur la physique mathématique*, Paris, Presses Universitaires de France, 1958. Edmund Husserl, *Logique formelle et logique transcendentale: Essai d'une critique de la raison logique*, trans. Suzanne Bachelard, Paris: Presses Universitaires de France, 1957. She also wrote an important study on this: Bachelard, *La logique de Husserl*, Paris: Presses Universitaires de France, 1957; Bachelard, *A Study of Husserl's Formal and Transcendental Logic*, trans. Lester E. Embree, Evanston, IL: Northwestern University Press, 1968.

43 Bachelard, 'Épistémologie et histoire des sciences', 49 n. 23.

44 Canguilhem draws on Jean-Toussaint Desanti, *La philosophie silencieuse, ou critique des philosophies de la science*, Paris: Seuil, 1975 (IR 17/7–8).

45 On progress, see 'La décadence de l'idée de progrès', *Revue de métaphysique et de morale* 92 (4), 1987, 437–54; 'The Decline of the Idea of Progress', trans. David Macey, *Economy and Society* 27 (2–3), 1998, 313–29.

46 Bogdan Suchodolski, 'Les facteurs du développement de l'histoire des sciences', in *XIIe Congrès international d'histoire des sciences*, 34.

8 Writings on Medicine

1 'Corps et santé', CAPHÉS GC 25.26, 21. See Geroulanos and Meyers, 'Introduction', WM −/8–9.
2 'Corps et santé', CAPHÉS GC 25.26, 22.
3 A different translation of this essay was published in 'Is a Pedagogy of Healing Possible?', trans. Steven Miller, *Umbr(a): Incurable*, 2006, 9–21.
4 *La santé, concept vulgaire et question philosophique*, Pin-Balma: Sables, 1998. The manuscript and related materials are archived at 'La santé, concept vulgaire et question philosophique, Strasbourg 1988, le 7 mai', CAPHÉS GC 24.5. An earlier version of the translation appeared as 'Health: Crude Concept and Philosophical Question', trans. Todd Meyers and Stefanos Geroulanos, *Public Culture*, 20 (3), 2008, 467–77.
5 Geroulanos and Meyers, 'Introduction', WM −/9. A fragment is preserved as CAPHÉS GC 25.29.
6 Georges Cuvier, *Histoire des progrès des sciences naturelles depuis 1789 jusqu'à ce jour*, 5 vols., Paris: Rouet, 1834, vol. I, 313–14; quoted in IR 47–8/41–2.
7 Canguilhem has Werner Leibbrand, *Die speculative Medizin der Romantik*, Hamburg: Claassen, 1956, in his reference list, but does not provide a page number.
8 John Brown, *The Elements of Medicine*, trans. the author, 2 vols., Portsmouth, NH: William & Daniel Treadwell, 1803 [1788], vol. I, 92, §CX.
9 Brown, *The Elements of Medicine*, vol. II, 285, §CCCXXVI.
10 Brown, *The Elements of Medicine*, vol. I, 242–3, §CCXLIV note; vol. II, 286, §CCCXXVII; vol. II, 278–9, §CCCXXII.
11 Canguilhem refers to NP 26–31/50–5 for a discussion of this 'ideological genealogy' (IR 54 n. 1 / 50 n. 12).
12 Epictetus, *Discourses*, Book II, §17.
13 Leriche, 'Équilibre de la santé et tempéraments', section 16, 1; quoted in WM 50/43. On Leriche, see also NP 67–77/91–101; 155–75/181–201.
14 Paul Valéry, *Mauvaises pensées et autres*, Paris: Gallimard, 1942, 191; quoted in WM 50/43.
15 Charles Daremberg, *La médecine, histoire et doctrines*, Paris: Didier et Cie, 1865, 323; quoted in WM 50/44.
16 Constitution of the World Health Organization, Principle I, www.who.int/governance/eb/who_constitution_en.pdf.

17 Xavier Bichat, *Recherches physiologiques sur la vie et la mort*, Verviers: Gérard & Co., 1973, 11; quoted in EHPS 74, 158, 225; and Canguilhem, 'De la science et de la contre-science', 179.
18 Immanuel Kant, 'The Conflict of the Faculties', trans. Mary J. Gregor, in *Religion and Rational Theology*, Cambridge University Press, 1996, 315. For a fuller discussion, see 'Therapeutique, Experimentation, Responsibilité' in EHPS, especially 386–9. In his text, Kant has theology, law and medicine as the superior faculties; while philosophy is in the inferior faculty. As Canguilhem notes elsewhere, philosophy here means 'letters and sciences'; but of the superior faculties, medicine is closest to philosophy (EHPS 386).
19 René Descartes to Hector Pierre Chanut, 31 March 1649 in *The Philosophical Writings of Descartes*, vol. III, 370, quoted in WM 53/45.
20 The translators note that in the 1977 Paris lecture 'Corps et santé', from which this 1988 text derives, Canguilhem added that Heidegger 'defines truth as nonveiling and exactitude' (WM – / 91 n. 13); see CAPHÉS GC 25.26, 2.
21 On Littré's role, generally, see Canguilhem, 'Émile Littré, philosophe de la biologie et de la médecine', in *Actes du colloque Émile Littré 1801–1881*, Paris: Albin Michel, 1983, 271–83.
22 Geroulanos and Meyers, WM – / 88 n. 30.
23 Ivan Illich, *Medical Nemesis: The Expropriation of Health*, New York: Pantheon, 1976; see *Deschooling Society*, Harmondsworth: Penguin, 1973.
24 On Illich, see also WM 93–4/63, which is not so critical, but suggests that it is hardly a new position.
25 Maurice Merleau-Ponty, *Le visible et l'invisible*, ed. Claude Lefort, Paris: Gallimard, 1964; Merleau-Ponty, *The Visible and the Invisible*, trans. Alphonso Lingis, Evanston, IL: Northwestern University Press, 1968; Merleau-Ponty, *L'union de l'âme et du corps chez Malebranche, Biran et Bergson*, ed. Jean Depruin, Paris: Vrin, 2nd edn, 1978 [1968]; Merleau-Ponty, *The Incarnate Subject: Malebranche, Biran, and Bergson on the Union of Body and Soul*, ed. Andrew Bjelland Jr and Patrick Burke, trans. Paul B. Milan, Amherst, NY: Humanity Books, 2001; Merleau-Ponty, *La Nature: Notes: Cours du Collège de France*, ed. Dominique Séglard, Paris: Seuil, 1996; translated by Robert Vallier as *Nature: Course Notes from the Collège de France*, Evanston, IL: Northwestern University Press, 2003.

26 Merleau-Ponty, *Le visible et l'invisible*, 283; Merleau-Ponty, *The Visible and the Invisible*, 234.

27 Merleau-Ponty, *Le visible et l'invisible*, 46–7; Merleau-Ponty, *The Visible and the Invisible*, 27.

28 As the translators note, 'The French term *expectant* has no simple English cognate'; having a sense of 'expectant', 'attentive' or 'wait-and-see'. See Geroulanos and Meyers, 'Introduction', WM –/22–3, – / 86 n. 4.

29 François Dagognet, *La raison et les remèdes*, Paris: Presses Universitaires de France, 1964. See also Dagognet, 'Nature naturante et nature dénaturée', in *Savoir, faire, espérer: Les limites de la raison*, Brussels: Facultés Universitaires Saint-Louis, vol. I, 1976, 71–87, and the lead essay in *Anatomie d'un épistémologue: François Dagognet*, Paris: Vrin, 1984.

30 Some of his references here (WM 31/33) include Max Neuburger, *The Doctrine of the Healing Power of Nature throughout the Course of Time*, trans. Linn J. Boyd, New York: [private printing], 1932; and Georg Groddeck, *Natura sanat, medicus curat: Der gesunde und kranke Mensch gemeinverständlich dargestellt*, Leipzig: Hirzel, 1913; translated by Pierre Villain, as *'Nasamecu': La nature guérit*, Paris: Aubier Montaigne, 1980. The title *Nasamecu* comes from the abbreviated phrase *'NAtura SAnat, MEdicus CUrat*; Nature heals, the doctor cures'.

31 The translation of the essay 'Diseases' restores three quotations from the manuscript, from Montaigne's *Essays*, Bernard's *Leçons sur le diabète et la glycogenèse animale* and Foucault's *Birth of the Clinic* (WM –/42); 'Les maladies', CAPHÉS GC 26.2.13, 18; cited in WM –/42 and 90 nn. 16, 17 and 18.

32 Denis Diderot, *Essais sur la peinture et Salons de 1759, 1761, 1763*, ed. Jacques Chouillet, Paris: Hermann, 1984, 11; Diderot, 'Notes on Painting', in *Diderot on Art: The Salon of 1765 and Notes on Painting*, ed. and trans. John Goodman, New Haven: Yale University Press, 1995, 191.

33 Jacques Tenon, *Mémoires sur les hôpitaux de Paris*, Paris: Ph. D. Pierres, 1788; *Memoirs on Paris Hospitals*, ed. and trans. Dora B. Weiner, Canton, MA: Science History Publications, 1996. This notion was the basis for a collaborative study: Michel Foucault, Blandine Barret Kriegel, Anne Thalamy, François Beguin and Bruno Fortier, *Les machines à guérir (aux origines de l'hôpital moderne)*, Paris: Institut de l'environnement, Bruxelles: Pierre Mardaga, revised edn, 1979 [1976]. Canguilhem reviewed the first edition in *Le Monde* 6 April 1977, 16. In 1978, Canguilhem

explicitly links this notion to the project led by Foucault in WM 86, 86 n. 2 / 60, 97 n. 23.

34   Bernard's notes, quoted in Léon Delhoume, 'Introduction', in Bernard, *Principes de médecine expérimentale*, xxix; cited in IR 64–5/60.

35   Both from Bernard, *Principes de médecine expérimentale*, 116, cited in IR 65/61.

36   François Dagognet, *Méthodes et doctrine dans l'oeuvre de Pasteur*, Paris: Presses Universitaires de France, 1967.

9   Legacies

1   Papers relating to the series are archived as 'Directeur de la Collection Hachette …', CAPHÉS GC 19.1. A letter from René Vaubourdolle at the press, 28 January 1952, suggests a number of possible titles for the series.

2   *Besoins et tendances*, texts selected and presented by G. Canguilhem, Paris: Hachette, 1952. References are to the version in OC IV.

3   *Introduction à l'histoire des sciences*, texts selected by S. Bachelard, J. C. Cadieux, G. Canguilhem et al., 2 vols., Paris: Hachette, 1970–1 – vol. I: *Éléments et instruments*; vol. II: *Objet, méthode, exemples*.

4   Camille Limoges, OC IV, 437 n. 1.

5   *Instincts et institutions*, texts selected and presented by G. Deleuze, Paris: Hachette, 1953; and *Sciences de la vie et de la culture*, texts selected and presented by F. Dagognet, Paris: Hachette, 1953.

6   Limoges, OC IV, 437 n. 2; *Technique et technologie*, texts selected and presented by Jacques Guillerme, Paris: Hachette, 1973.

7   The manuscript of this text, dated to 10 November 1951, is archived as CAPHÉS GC 19.1.1.

8   'Avant-Propos', *Introduction à l'histoire des sciences*, vol. I, iii–v.

9   'Avant-Propos', iii.

10   'Avant-Propos', iv.

11   'Avant-Propos', iv.

12   Aside from the works already referenced, see François Dagognet, *Le catalogue de la vie: Étude méthodologique sur la taxinomie*, Paris: Presses Universitaires de France, 1970; Jean-Charles Sournia, *Mythologies de la médecine moderne: Essai sur le corps et la raison*, Paris: Presses Universitaires de France, 1969.

13 Paracelse, *Oeuvres médicales*, ed. and trans. Bernard Gorceix, Paris: Presses Universitaires de France, 1968; Yvette Conry, *Correspondance entre Charles Darwin et Gaston de Saporta*, Paris: Presses Universitaires de France, 1972.

14 Evelyne Aziza-Shuster, *Le médecin de soi-même*, Paris: Presses Universitaires de France, 1972.

15 See also Michel Morange, 'Georges Canguilhem et la biologie du XXe siècle', *Revue d'histoire des sciences* 53 (1), 2000, 83–106; and his *A History of Molecular Biology*, trans. Matthew Cobb, Cambridge, MA: Harvard University Press, 1998. For reconsiderations, see Jean Gayon, 'The Concept of Individuality in Canguilhem's Philosophy of Biology', *Journal of the History of Biology* 31 (3), 1998, 305–25; Jonathan Hodge, 'Canguilhem and the History of Biology', *Revue d'histoire des sciences* 53 (1), 2000, 65–82.

16 Paul Rabinow, 'Foreword', in François Delaporte, *Disease and Civilization: The Cholera in Paris, 1832*, trans. Arthur Goldhammer, Cambridge, MA: MIT Press, 1986, x.

17 Rabinow, 'Foreword', x–xi. See also Rabinow, 'French Enlightenment: Truth and Life', *Economy and Society* 27 (2–3), 1998, 193–201.

18 Ragon, 'King Cang', 20.

19 Rabinow, 'Foreword', xi.

20 'Introduction: La Constitution de la physiologie comme science', in Charles Kayser et al., *Physiologie*, 3 vols., Paris: Éditions Medicales Flammarion, 1963, vol. I, 11–48; this copy is archived as Foucault 1430, Beinecke Rare Book and Manuscript Library, Yale University.

21 Foucault's preparatory notes for this course are archived at the Bibliothèque nationale de France, NAF 28730, Boxes 39 and 45, published as *La sexualité: Cours donné à l'université de Clermont-Ferrand (1964), suivi de Le discours de sexualité*, ed. Claude-Olivier Doran, Paris: Seuil/Gallimard, 2018.

22 Foucault, *Dits et écrits*, vol. I, 844–5; Foucault, *Essential Works*, vol. I, 7–8. See Foucault, *L'ordre du discours*, 33–8, 70–1; Foucault, 'The Order of Discourse', 60–1, 72–3. See also Foucault's review of Jacob's *The Logic of Life* in *Dits et écrits*, vol. II, 99–104.

23 Foucault, *Dits et écrits*, vol. III, 210; Foucault, *Essential Works*, vol. III, 137.

24 See, for example, Michel Foucault, *Histoire de la sexualité I: La Volonté de savoir*, Paris: Gallimard/Tel, 1976, 73; translated by

Robert Hurley as *The History of Sexuality I: The Will to Knowledge*, London: Penguin, 1978, 54–5.

25 The references to all these issues would be to almost all of Foucault's writings and lectures. For a much fuller discussion, see Stuart Elden, *Foucault: The Birth of Power*, Cambridge: Polity, 2017, especially ch. 7; and Elden, *Foucault's Last Decade*.

26 François Delaporte, *Le second règne de la nature: essai sur les questions de la végétalité au XVIIIe siècle*, Paris: Flammarion, 1979; Delaporte, *Nature's Second Kingdom*, trans. Arthur Goldhammer, Cambridge, MA: MIT Press, 1982. Canguilhem's preface is on 7–10/ix–xii.

27 Delaporte, *Disease and Civilization*; revised French version as *Le savoir de la maladie*, Paris: Presses Universitaires de France, 1990; Delaporte, *Histoire de la fièvre jaune: Naissance de la médecine tropicale*, Paris: Payot, 1989; Delaporte, *The History of Yellow Fever*, trans. Arthur Goldhammer, Cambridge, MA: MIT Press, 1989; Delaporte, *La Maladie de Chagas: Histoire d'un fléau continental*, Paris: Payot, 1999; Delaporte, *Chagas Disease: History of a Continent's Scourge*, trans. Arthur Goldhammer, New York: Fordham University Press, 2012; Delaporte, *Les épidémies*, Paris: Presses Pocket, 1995; Delaporte, *Histoire des myopathies* (with Patrice Pinell), Paris: Payot, 1998. See also Delaporte, 'The Discovery of the Vector of Robles Disease', *Parassitologia* 50 (3–4), 2008, 227–31.

28 François Delaporte, *Anatomie des passions*, Paris: Presses Universitaires de France, 2003; Delaporte, *Anatomy of the Passions*, trans. Susan Emanuel, Stanford University Press, 2008; Delaporte, *Figures de la médecine*, Paris: Éditions de Cerf, 2009; Delaporte, *Figures of Medicine: Blood, Face Transplants, Parasites*, trans. Nils F. Schott, New York: Fordham University Press, 2013.

29 Delaporte, *Le savoir de la maladie*, 5; Delaporte, *Disease and Civilization*, xvii.

30 Delaporte, *Histoire de la fièvre jaune*, 7; Delaporte, *The History of Yellow Fever*, xiii.

31 Delaporte, *Disease and Civilization*, 6 (not in French).

32 Rabinow, 'Foreword', ix.

33 Delaporte, *La maladie de Chagas*, 18; Delaporte, *Chagas Disease*, 9.

34 See Todd Meyers, 'Foreword: The False Cannot Be the Moment of the True', in Delaporte, *Chagas Disease*, ix–x. For an updating of his study to look at the present, see Delaporte, *Figures of Medicine*, ch. 6.

35 'Préface', in *Le second règne de la nature*, 9; 'Foreword', in *Nature's Second Kingdom*, xi.
36 'Préface', 9; 'Foreword', xi.
37 'Présentation', in Delaporte, *Histoire de la fièvre jaune*, 12; 'Foreword', in Delaporte, *The History of Yellow Fever*, x.
38 Delaporte, *La maladie de Chagas*, 14; Delaporte, *Chagas Disease*, 6.
39 Delaporte, *La maladie de Chagas*, 17; Delaporte, *Chagas Disease*, 8. Foucault, *Dits et écrits*, vol. III, 14; Foucault, *Essential Works*, vol. III, 91. For a discussion of the collaborative work that this developed from, see Elden, *Foucault: The Birth of Power*, ch. 6; and Elden, *Foucault's Last Decade*, ch. 4.
40 'Présentation', in Delaporte, *Histoire de la fièvre jaune*, 12; 'Foreword', in Delaporte, *The History of Yellow Fever*, x.
41 'Présentation', in Delaporte, *Histoire de la fièvre jaune*, 12–13; 'Foreword', in Delaporte, *The History of Yellow Fever*, x–xi.
42 Foucault, *Dits et écrits*, vol. IV, 763–4; Foucault, *Essential Works*, vol. II, 466. The original version of this text makes the link to Althusser and Althusserians explicit (*Dits et écrits*, vol. III, 429; NP –/8).
43 In relation to contemporary medical research, see Annemarie Mol, 'Lived Reality and the Multiplicity of Norms: A Critical Tribute to George Canguilhem', *Economy and Society* 27 (2–3), 1998, 274–84.
44 Foucault, *Dits et écrits*, vol. IV, 435; Foucault, *Essential Works*, vol. II, 437.
45 Bruno Latour and Steve Woolgar, *Laboratory Life: The Social Construction of Scientific Facts*, London: Sage, 1979; Bruno Latour, *Science in Action: How to Follow Scientists and Engineers through Society*, Cambridge, MA: Harvard University Press, 1987.
46 See Lorna Weir, 'Cultural Intertexts and Scientific Rationality: The Case of Pregnancy Ultrasound', *Economy and Society* 27 (2), 1998, 249–58.
47 Ian Hacking, 'Canguilhem Amid the Cyborgs', *Economy and Society* 27 (2–3), 1998, 202–16; Donna Jeanne Haraway, *Crystals, Fabrics and Fields: Metaphors of Organicism in Twentieth-century Developmental Biology*, New Haven: Yale University Press, 1976.
48 See, for example, Chris Philo, 'A Vitally Human Medical Geography? Introducing Georges Canguilhem to Geographers', *New Zealand Geographer* 63, 2007, 82–96; S. M. Reid-Henry, *The Cuban Cure: Reason and Resistance in Global Science*, University of Chicago Press, 2010.

49   Bernard Stiegler, *Automatic Society Volume 1: The Future of Work*, trans. Daniel Ross, Cambridge: Polity, 2017.

50   Catherine Malabou, *Métamorphoses de l'intelligence: Que faire de leur cerveau bleu?* Paris: Presses Universitaires de France, 2017, 63–7.

51   Braunstein, 'Présentation', 10.

52   Foucault, *Dits et écrits*, vol. IV, 764; Foucault, *Essential Works*, vol. II, 466.

53   Foucault, *Dits et écrits*, vol. IV, 767; Foucault, *Essential Works*, vol. II, 469.

54   We might also note Canguilhem's use of Merleau-Ponty's work, notably *The Structure of Behaviour*.

55   Foucault, *Dits et écrits*, vol. IV, 773–4; Foucault, *Essential Works*, vol. II, 475.

56   Foucault, *Dits et écrits*, vol. IV, 776; Foucault, *Essential Works*, vol. II, 477.

57   'Qu'est-ce qu'un philosophe en France aujourd'hui?' *Commentaire* 53 (1), 1991, 107–12. See Lecourt, *Georges Canguilhem*, 99–100.

Index